T0289125

Optical Sensors

An introduction with lab demonstrations

Online at: https://doi.org/10.1088/978-0-7503-4876-8

IOP Series in Emerging Technologies in Optics and Photonics

Series Editor

R Barry Johnson, a Senior Research Professor at Alabama A&M University, has been involved for over 50 years in lens design, optical systems design, electro-optical systems engineering, and photonics. He has been a faculty member at three academic institutions engaged in optics education and research, has been employed by a number of companies, and has provided consulting services.

Dr Johnson is an IOP Fellow, an SPIE Fellow and Life Member, an OSA Fellow, and was the 1987 President of SPIE. He serves on the editorial board of *Infrared Physics & Technology* and *Advances in Optical Technologies*. Dr Johnson has been awarded many patents, has published numerous papers and several books and book chapters, and was awarded the 2012 OSA/SPIE Joseph W Goodman Book Writing Award for Lens Design Fundamentals (second edition). He is a perennial co-chair of the annual SPIE Current Developments in Lens Design and Optical Engineering Conference.

Foreword

Until the 1960s the field of optics was primarily concentrated in the classical areas of photography, cameras, binoculars, telescopes, spectrometers, colorimeters, radiometers, etc. In the late 1960s optics began to blossom with the advent of new types of infrared detector, liquid crystal display (LCDs), light emitting diode (LEDs), charge coupled device (CCDs), laser, holography, and fiber optics along with new optical materials, advances in optical and mechanical fabrication, new optical design programs, and many more technologies. With the development of the LED, LCD, CCD, and other electro-optical devices, the term 'photonics' came into vogue in the 1980s to describe the science of using light in the development of new technologies and the operation of a myriad of applications. Today optics and photonics are truly pervasive throughout society and new technologies are continuing to emerge. The objective of this series is to provide students, researchers, and those who enjoy self-education with a wide-ranging collection of books, each of which focuses on a topic relevant to the technologies and applications of optics and photonics. These books will provide knowledge to prepare the reader to be better able to participate in these exciting areas now and in the future. The title of this series is *Emerging Technologies in Optics and Photonics*, in which 'emerging' is taken to mean 'coming into existence', 'coming into maturity', and 'coming into prominence'. IOP Publishing and I hope that you will find this series of significant value to you and your career.

A full list of titles published in this series can be found here: https://iopscience.iop.org/bookListInfo/emerging-technologies-in-optics-and-photonics.

Optical Sensors

An introduction with lab demonstrations

Victor Argueta-Diaz

Physics and Engineering Department, Alma College (United States), Alma, MI, USA

IOP Publishing, Bristol, UK

ISBN 978-0-7503-4876-8 (ebook)
ISBN 978-0-7503-4874-4 (print)
ISBN 978-0-7503-4877-5 (myPrint)
ISBN 978-0-7503-4875-1 (mobi)

DOI 10.1088/978-0-7503-4876-8

Version: 20230801

IOP ebooks

British Library Cataloguing-in-Publication Data: A catalogue record for this book is available from the British Library.

Published by IOP Publishing, wholly owned by The Institute of Physics, London

IOP Publishing, No.2 The Distillery, Glassfields, Avon Street, Bristol, BS2 0GR, UK

US Office: IOP Publishing, Inc., 190 North Independence Mall West, Suite 601, Philadelphia, PA 19106, USA

To my wife and kids, I love you to the Moon and back.

To my parents and sisters. I miss you all so much.

—VA

Contents

Preface

Optics is a fascinating field that has been studied for centuries, and it continues to be an area of great interest and innovation today. From the earliest investigations of light and vision to the latest breakthroughs in laser technology, optics has played a critical role in advancing our understanding of the natural world and shaping our technological capabilities. Optical sensors are an important class of sensors that are very attractive due to their versatility, accuracy, and non-invasive nature. They have found applications in a wide range of fields, from biomedical imaging to environmental monitoring and beyond. This book provides an introduction to the field of optical sensors, covering the principles of operation, design, and applications of different types of optical sensors.

This book is an introduction to the principles of optics, designed for students and professionals who are interested in learning more about this exciting field. Throughout its pages, you will find a comprehensive overview of the fundamental concepts of optics, including topics such as Maxwell's equations, wave optics, polarization, interference, and diffraction.

One of the main goals of this book is to provide a clear and accessible explanation of the key concepts and techniques used in optics while also demonstrating their practical applications. Whether you are interested in designing optical systems, conducting research in the field of optics, or simply learning more about this fascinating subject, this book will provide you with a solid foundation upon which to build your knowledge and skills.

The book also has a series of lab experiments that, I hope, can be implemented without a major investment from what a properly equipped college optics lab may already have. The labs are designed to teach some basic principles about optical sensors and how to handle optical equipment properly.

This book is designed to be accessible to both students and professionals without the need to have vast experience in optics, however, it requires knowledge of vector calculus and differential equations. I hope this book will provide a comprehensive introduction to the field of optical sensors, making it an invaluable resource for researchers, engineers, and scientists working in the field.

—VA

Acknowledgments

Writing this book has been a long and challenging process, and I'm immensely grateful to all the individuals who helped me finish this work.

First, I would like to thank my wife, your love and support have been a source of strength and inspiration to me, and I am forever grateful for all that you do.

Thank you for your beautiful smile, for always having my back, and for helping me build this beautiful family of ours. You are an amazing mother, and your love and guidance have helped them grow into kind, compassionate, and intelligent individuals.

Leo, I wanted to take a moment to express how proud I am of you and the person you are becoming. Watching you grow and develop into such an amazing, insightful man has been a true joy and honor.

Your creativity, passion, and dedication to your art have been truly inspiring, and I am grateful for all the help you have given me with it. Your eye for detail, your ability to see the beauty in the world, and your commitment to your craft are all qualities that will serve you well throughout your life.

Rowan, I wanted to take a moment to express my appreciation for your amazing positivity and for being such a wonderful horror-movie-watching partner.

Your enthusiasm for life and ability to see the best in every situation are truly inspiring. You bring so much light and joy into my life, and I am grateful for the moments we share together. Whether we watch scary movies or spend time together, your presence always fills me with a sense of happiness and contentment. I'm so happy you are not an alien!

I am sincerely grateful to the publishers, editors, and production staff who labored assiduously to bring this book to life. Your diligence and professionalism have rendered the publication process effortless and delightful.

I would like to thank my students, who have encouraged and supported me throughout this project. Your enthusiasm has been humbling.

—VA

Author biography

Victor Argueta-Diaz

Figure 1: The author pretending to casually be working while someone has a camera three inches from his face.

(Born Mexico City, 1974) I received my BS degree in Telecommunication Engineering from the National Autonomous University of Mexico, in 1999. My MS degree in Electrical and Computer Engineering in 2002, and my PhD degree in Optoelectronics from The Ohio State University in 2005. While at OSU, I was granted six patents in the area of optical communications.

After doing a postdoc back in Mexico, developing optical sensors to detect air pollutants and 3D imaging, I returned to the US, where I worked in the biomedical industry, developing software for a couple of years.

Since 2012, I have been a professor of physics and engineering at Alma College, where I try to find the intersections among science, arts, and humanities.

Part I

Basic principles and components

Chapter 1

Introduction

Optical sensors have become ubiquitous in modern technology, finding applications in a variety of fields such as medical diagnosis, environmental monitoring, industrial automation, and consumer electronics. Optical sensors rely on the interaction of light with materials to detect changes in the physical, chemical, or biological environment. The sensitivity, selectivity, and accuracy of optical sensors make them particularly useful for detecting low concentrations of analytes in complex matrices.

Optical sensing is based on the interaction between light and matter. When light is absorbed, reflected, or scattered by an object, it can provide information about the object's properties such as its composition, shape, size, and distance. Optical sensors use various techniques to detect and measure these properties. Some common techniques are:

- **Absorption:** this technique measures the amount of light absorbed by a material. It is used to detect the presence of certain gases and liquids, such as carbon monoxide and glucose, that have specific absorption spectra.
- **Reflection:** this technique measures the amount of light reflected by a surface. It is used to detect the position and movement of objects, such as in optical encoders and proximity sensors.
- **Scattering:** this technique measures the amount of light scattered by a material. It is used to detect the size and concentration of particles in fluids, such as in laser particle counters and turbidity meters.
- **Interference:** this technique measures the phase shift and amplitude change of light waves when they pass through a material. It is used to measure thickness, refractive index, and surface roughness, such as in interferometric sensors.

1.1 History

Many ancient Greek philosophers, such as Euclid and Ptolemy, made substantial contributions to the field of optics throughout their careers. However, the first systematic investigations into optics started in the 17th century, when the scientific

doi:10.1088/978-0-7503-4876-8ch1

method was first developed. Over the course of several centuries, numerous important discoveries and advancements have been achieved in the field of optics. These advancements include the creation of the light wave theory, the construction of the telescope and microscope, and the growth of modern optics. An outline of optics development is provided here in a condensed form.

1.1.1 17th Century

During the 17th century, significant advancements were made in the field of optics, which had a profound impact on the way we understand light and vision. One of the most important contributions to optics during this period was the development of the telescope by Dutch lens grinders [3, 8], which allowed for new observations of the heavens. This enabled Galileo Galilei to make observations that contradicted traditional cosmological assumptions, thereby challenging long-held beliefs about the nature of the Universe.

Another notable contribution to the field was made by Sir Isaac Newton, who proposed the theory of colors that challenged the traditional view that colors were the result of the modification of white light. As noted by [7], Newton's work on the nature of light and color was a fundamental contribution to the field of optics that revolutionized our understanding of how light behaves and how we perceive color.

The wave theory of light was also introduced during this period by Huygens, who proposed that light travels in waves rather than particles, as previously believed. According to [5], this theory provided a new framework for understanding light and laid the foundation for the study of optics in the centuries to come.

The 17th century was a period of significant advancements in the field of optics, with notable contributions made by Galileo Galilei, Sir Isaac Newton, and Huygens. These contributions challenged traditional beliefs and laid the foundation for further study in the field of optics.

1.1.2 18th Century

One of the most important advances in optics in the 18th century was the development of the first practical microscopes. In 1759, John Dollond invented the achromatic lens, which corrected for chromatic aberration, a defect that had previously limited the usefulness of microscopes [5]. This allowed for the production of much sharper images, and it opened up a new world of microscopic detail for scientists to explore.

Another important advance in optics in the 18th century was the development of the first practical telescopes. In 1704, Isaac Newton invented the reflecting telescope, which used a mirror to collect and focus light [5, 13]. This was a much more powerful design than the refracting telescopes that had been used previously, and it allowed astronomers to see objects that were much fainter and further away.

The discovery of polarization was also a major advance in optics in the 18th century [6]. In 1784, Etienne Malus discovered that light could be polarized, meaning that its waves could be aligned in a specific direction. This discovery led to the development of new optical devices, such as polarizing filters, which are used in a variety of applications, from sunglasses to 3D televisions.

The wave theory of light was also developed in the 18th century [5]. In 1678, Christiaan Huygens proposed that light travels in waves. This proposal was supported by the experiments of Thomas Young in 1801, who showed that light can interfere with itself. The wave theory of light eventually replaced the particle theory of light, which had been dominant for centuries.

1.1.3 19th Century

One of the most significant advances in optics during the 19th century was the wave theory of light. Before this, it was believed that light traveled in a straight line, but Thomas Young and Augustin-Jean Fresnel proposed that light behaved like a wave, which helped to explain a wide range of optical phenomena, including diffraction and interference [5]. This new theory provided a more comprehensive understanding of the behavior of light and paved the way for many other discoveries in optics.

Another significant development in optics during the 19th century was the emergence of spectroscopy, which is the study of the interaction between light and matter. Gustav Kirchhoff and Robert Bunsen used spectroscopy to discover new elements and develop new methods of chemical analysis [5, 8, 15]. They found that when light passes through a sample of material, it is absorbed by certain wavelengths, which can be used to identify the presence of specific elements in the sample. This discovery had important implications for chemistry and the study of the natural world.

Photography was also invented during the 19th century, which revolutionized the way we see and record the world. Joseph Nicéphore Niépce and Louis Daguerre invented photography, paving the way for new forms of art, science, and communication [12]. The ability to capture images on film opened up new avenues for scientific inquiry, artistic expression, and personal documentation. Photography has since become an essential part of modern life and is used in many areas, including science, art, journalism, and advertising.

The 19th century saw the development of new optical instruments, such as the microscope and telescope, which made it possible to see and study objects at a much greater level of detail [14]. This led to many important discoveries in fields such as biology, astronomy, and physics. The microscope, in particular, allowed scientists to study the structure of cells and microorganisms, leading to new insights into the workings of the natural world.

Polarization was also a significant area of study in optics during the 19th century [6]. Étienne-Louis Malus and William Nicol pioneered the study of polarized light, which helped to explain many optical phenomena, including the colors seen in polarized sunglasses and the structure of crystals. This discovery had implications for both science and technology, as polarized light is used in many areas, including microscopy, astronomy, and optical communication.

Finally, the 19th century saw the emergence of James Clerk Maxwell's theory of electromagnetism, which showed that light is an electromagnetic wave [1]. This discovery helped to unify the study of electricity, magnetism, and optics and provided a more comprehensive understanding of the nature of light.

1.1.4 20th Century

One of the most important advances in optics in the 20th century was the development of fiber optics. In 1966, Charles Kao and George Hockham theorized that glass fibers could be used to transmit light signals over long distances with little loss [11].

Kao and Hockham's work was met with skepticism at first, but they eventually convinced the Standard Telecommunications Laboratories (STC) to fund their research. In 1966, they published a paper in the journal *Electronics Letters* that outlined their theory and the results of their experiments. This paper is considered to be the seminal work on fiber optics [11].

In the years that followed, Kao and Hockham continued to work on improving the performance of fiber optic cables. In 1970, they were joined by Robert Maurer, Donald Keck, and Peter Schultz, a team of scientists working for Corning Glass Works. This team was able to develop a method for manufacturing glass fibers with very low loss.

In 1977, Corning Glass Works announced the commercial availability of fiber optic cables. These cables were used in the first commercial fiber optic telecommunications system, which was installed in Chicago in 1980 [2, 10].

Fiber optics has since become an essential part of our telecommunications infrastructure. It is used to transmit data, voice, and video signals over long distances with high speed and reliability. Fiber optics is also used in a variety of other applications, including medical imaging, industrial inspection, and military communications.

The first laser was invented in 1960 by Theodore Maiman at Hughes Research Laboratories. Maiman's laser was a ruby laser, and it was based on the theoretical work of Charles H Townes and Arthur L Schawlow [4, 9, 16].

Townes and Schawlow had proposed the idea of a laser in 1958, and they had received a patent for the idea in 1960. However, Maiman was the first actually to build a working laser [16]. He did this by using a ruby crystal as the lasing medium. He pumped the ruby crystal with a flash lamp, and he was able to produce a beam of laser light.

Maiman's invention of the laser was a major breakthrough in the field of optics. Lasers have since become an essential part of our technology, and they are used in a wide variety of applications. Some of the most common applications of lasers include: telecommunications, medicine, manufacturing and entertainment to mention a few.

In the 20th century, the development of lasers and fiber optics led to a new generation of optical sensors [4, 9, 16]. These sensors are more sensitive and accurate than earlier types of sensors, and they can be used to measure a wider range of physical quantities, optical sensors can be classified into several categories based on their working principles, configurations, and applications. Some common types of optical sensors are:

- **Fiber optic sensors:** these sensors use optical fibers to transmit and receive light signals. They can be configured as point sensors or distributed sensors, depending on the sensing element and the readout method. Fiber optic sensors have many advantages, such as immunity to electromagnetic interference, high sensitivity, and multiplexing capability. They are used in various applications, such as structural health monitoring, temperature sensing, and chemical sensing.
- **Photonic sensors:** these sensors use photonic devices, such as photodiodes, phototransistors, and photovoltaic cells, to convert light signals into electrical signals. They can be configured as discrete or integrated sensors, depending on the device structure and the circuit design. Photonic sensors have many advantages, such as fast response, high resolution, and low noise. They are used in various applications, such as optical communication, imaging, and spectroscopy.
- **Surface plasmon sensors:** these sensors use surface plasmon resonance (SPR) to measure the refractive index and thickness of thin films and biomolecules. They can be configured as prism-based or grating-based sensors, depending on the SPR coupling mechanism and the readout method. Surface plasmon sensors have many advantages, such as label-free detection, high sensitivity, and real-time monitoring. They are used in various applications, such as biosensing, drug discovery, and environmental monitoring.
- **Micro-electromechanical systems (MEMS) sensors:** these sensors use microfabrication techniques to create optical components and structures on a small scale. They can be configured as accelerometers, gyroscopes, pressure sensors, and microfluidic sensors, depending on the MEMS design and the sensing mechanism. MEMS sensors have many advantages, such as low power consumption, small size, and high reliability. They are used in various applications, such as consumer electronics, medical devices, and automotive systems.

1.2 Growth expectations

The global market for optical sensors is expected to grow at a compound annual growth rate (CAGR) of 11% from 2021 to 2028. The growth of the market is attributed to the increasing demand for optical sensors in various end-use industries, such as industrial, medical, automotive, and consumer electronics.

The industrial segment is expected to be the largest market for optical sensors during the forecast period. The growth of this segment is attributed to the increasing use of optical sensors in industrial process control, quality control, and safety applications. The medical segment is also expected to grow at a significant rate during the forecast period. The growth of this segment is attributed to the increasing use of optical sensors in medical diagnostics, surgical procedures, and patient monitoring applications.

The automotive segment is expected to grow at a moderate rate during the forecast period. The growth of this segment is attributed to the increasing use of optical sensors in advanced driver assistance systems (ADAS), autonomous driving, and vehicle safety applications. The consumer electronics segment is expected to grow at a high rate during the forecast period. The growth of this segment is attributed to the increasing use of optical sensors in smartphones, tablets, laptops, and other consumer electronic devices.

The key players operating in the global market for optical sensors are Honeywell International Inc., Infineon Technologies AG, Omron Corporation, Panasonic Corporation, Robert Bosch GmbH, STMicroelectronics, Texas Instruments Incorporated, and Vishay Intertechnology Inc. These players are focusing on expanding their product portfolio and geographical presence to gain a competitive edge in the market.

The future of optical sensors is very promising. New developments in materials science and nanotechnology are leading to the development of even more sensitive and accurate optical sensors. These sensors will be used in a wide range of new applications, such as: self-driving cars, biometric identification, and early cancer detection.

1.3 Book overview

The book covers a broad range of topics related to optical sensors. The first part begins with a discussion of light sources and detectors, including the various types of light sources that are commonly used in optical systems, as well as the different types of detectors that are used to measure light.

The book then delves into the underlying theoretical framework of optics, starting with an overview of Maxwell's equations, which describe the behavior of electromagnetic waves. The book explains the mathematical basis of these equations and how they are used to model light propagation in various optical systems.

The book then moves on to discuss electromagnetic theory, including the behavior of light in various materials and the principles of reflection and refraction. The book covers topics such as polarization, dispersion, and the interaction of light with matter.

The book also covers physical optics, which is concerned with the wave-like nature of light and the ways in which it interacts with matter. This includes topics such as interference, diffraction, and the propagation of light through optical media.

Finally, the book covers diffraction and optical waveguides, it covers the fundamentals of optical waveguides, including their structure, properties of guided modes, and operation.

The book's second part is lab demonstrations that help us demonstrate some of the implementations of optical sensors. We start with some basic procedures to perform laser alignment and beam characterization measurements. We then describe some experiments to measure the refractive index and attenuation coefficients of liquids.

We then work with some experiments to control the polarization of light and detect the Brewster angle, followed by a couple of interferometer configurations

(Michelson and Fabry–Perot) and evaluate the high sensitivity that we can achieve with this system. In particular, we are interested in measuring the wavelength of different sources, wavelength resolution on a multi-wavelength source, and how to measure the refractive index of some solids.

Finally, we will do some experiments involving diffraction, both Fraunhofer and Fresnel. We will be observing the patterns formed by the diffracted light by different kinds of apertures. We will then use this experience to design an optical spectrometer.

The third and last part of the book covers some applications of optical sensors, we describe the principle of operation of Light Detection and Ranging (LiDAR) technology, which is an active remote sensing system that uses lasers to measure the distance between the sensor and a target. LiDAR has a range of applications, including 3D mapping, environmental monitoring, and autonomous vehicles.

We then do a review of optical sensors. Optical biosensors are used in various applications such as medical diagnostics, food safety, and environmental monitoring. We will describe the principles of operation, their classification, and some basic applications.

Overall, the book provides a comprehensive overview of optical sensors, covering both the theoretical principles of the subject and its practical applications in a range of fields. The text is accessible to readers of all levels, and includes examples and figures to help illustrate the points being made. Overall, this is a valuable resource for researchers, engineers, and students alike.

References

[1] Balanis C A 2012 *Advanced Engineering Electromagnetics* 2nd edn (New York: Wiley)

[2] Ballato J and Dragic P 2016 Glass: the carrier of light–a brief history of optical fiber *Int. J. Appl. Glass Sci.* **7** 413–22

[3] Bedini S A 1966 Lens making for scientific instrumentation in the seventeenth century *Appl. Opt.* **5** 687–94

[4] Bertolotti M 2004 *The History of the Laser* (Boca Raton, FL: CRC Press)

[5] Darrigol O 2012 *A History of Optics from Greek Antiquity to the Nineteenth Century* (Oxford: Oxford University Press)

[6] Goldstein D H 2017 *Polarized Light* (Boca Raton, FL: CRC Press)

[7] Gross A G 1988 On the shoulders of giants: seventeenth-century optics as an argument field *Q. J. Speech* **74** 1–17

[8] Hecht J 1991 *City of Light: The Story of Fiber Optics* (Oxford: Oxford University Press)

[9] Hecht J 2010 Short history of laser development *Opt. Eng.* **49** 091002

[10] Kanamori H 2020 Fifty year history of optical fibers *SEI Tech. Rev.* **91** 15–22

[11] Charles Kao K and Hockham George A 1966 Dielectric-fibre surface waveguides for optical frequencies *Proc. Inst. Electr. Eng.* **113** 1151–8

[12] Kingslake R 1992 Online access with subscription: SPIE Digital Library *Optics in Photography* (Bellingham, WA: SPIE Optical Engineering Press)

[13] Pedrotti F L, Pedrotti L M and Pedrotti L S 2017 *Introduction to Optics* (Cambridge: Cambridge University Press)

[14] Smith G and Atchison D A 1997 *The Eye and Visual Optical Instruments* (Cambridge: Cambridge University Press)
[15] Tennyson J 2019 *Astronomical Spectroscopy: An Introduction to the Atomic and Molecular Physics of Astronomical Spectroscopy* 3rd edn (London: World Scientific Publishing)
[16] Thyagarajan K and Ghatak A 2010 *Lasers: fundamentals and Applications* (Berlin: Springer Science)

IOP Publishing

Optical Sensors
An introduction with lab demonstrations
Victor Argueta-Diaz

Chapter 2

Light sources and detectors

An optical sensor is a device that converts light into an electrical signal. The light may be visible, infrared, or ultraviolet. Optical sensors are used in a variety of applications, including automotive lighting, optical navigation, medical diagnostics, and machine vision.

The choice of the correct light source is essential in the design and implementation of an optical sensor. The quality and optical characteristic of the light source in many cases defines the resolution and limitations of our sensors.

When choosing an optical source, it is important to consider its emissions wavelength, coherence, emission power, and polarization. These characteristics can have a significant impact on the quality of the light, as well as its suitability for certain applications.

The emissions wavelength of a light source is the wavelength of the light it emits. This is an important consideration because the wavelength of light can affect how it is detected and measured. For example, some sensors are only able to detect light within a certain wavelength range. In addition, the wavelength of light can also affect its propagation through different materials. For instance, optical fibers are optimized to work at 850 nm or 1550 nm.

The coherence of a light source is an indicator of how closely its radiations are related in time. This has a big impact on the perceptibility of disturbance patterns, making it preferable to have a higher degree of coherence to enhance the light's quality.

The emission power of a light source is the power of the light it emits. This is an important consideration because it can affect the brightness of the light. In general, it is desirable for a light source to have a high emission power.

In this chapter we will briefly explain some of these optical characteristics, we will describe some of the most common light sources used in optical sensors, and we will present some general guidelines on how to choose the correct light source for a specific application.

doi:10.1088/978-0-7503-4876-8ch2

2.1 Optical properties of light sources

2.1.1 Emission wavelength

An important optical characteristic of any light source is the emission wavelength(s). Natural light sources tend to have a broad emission covering a large range of wavelengths. For example, sunlight covers a range of 250 nm (deep in the ultra-violate) all the way to 2500 nm, however, most of its power is concentrated in the visible range between 400 nm and 750 nm, with a peak around 500 nm as shown in figure 2.1. The Sun produces light with a distribution similar to what would be expected from a 5525 K (5250 °C) blackbody, which is approximately the Sun's surface temperature.

When artificially creating light sources, particularly for specific purposes, it is often possible to engineer the emission to be much narrower. For example, a mercury-vapor lamp used in microscopy may be designed to emit light mostly in the blue and green regions, to better illuminate specimens for human observers. In this case, the lamp would be said to have 'peak emission' in the blue–green region of the visible spectrum.

The emission wavelength(s) of a light source are important to consider for many reasons. One reason is that the human eye is most sensitive to light in the green region of the visible spectrum, so a lamp with peak emission in the green will appear significantly brighter than one with peak emission in the blue or red. This is why sodium-vapor lamps, which have peak emission in the yellow–orange, are often used

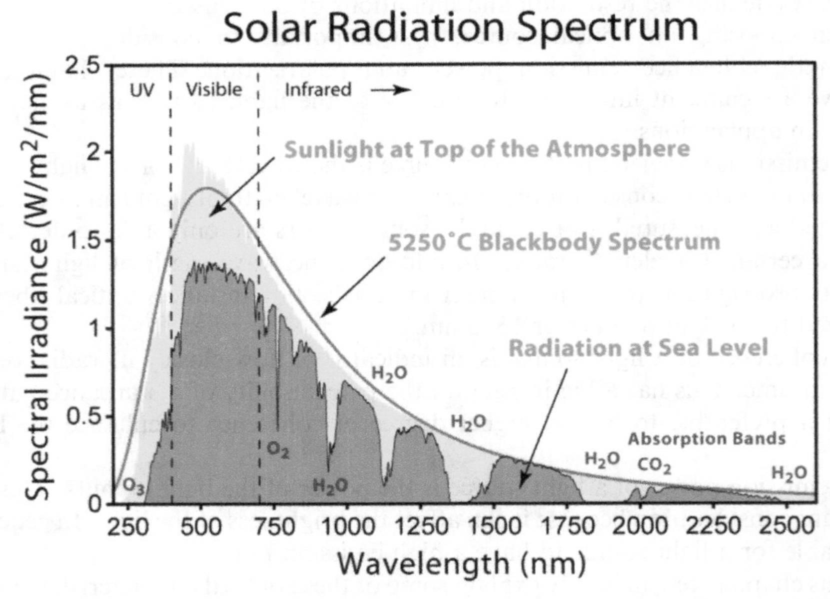

Figure 2.1. Solar radiation spectrum for direct light at both the top of the Earth's atmosphere (represented by the area in yellow) and at sea level (area in red). Image and description from [12] has been obtained by the author from the Wikimedia website where it was made available by User:Dragons flight under a CC BY-SA 3.0 licence. It is included within this article on that basis. It is attributed to User:Dragons flight.

for street lighting—they appear much brighter to human observers than lamps with peak emission in other regions.

Another reason to consider emission wavelength(s) is that different materials will absorb and reflect light of different wavelengths. For example, when working with fluorescent microscopes, organic samples may absorb light in the ultraviolet region while emitting in the visible region.

The emission wavelength of a light source is a measure of the spectral composition of the visible light emitted by that source. The emission wavelength tells us what colors are emitted by the source, and how much power is emitted at each wavelength.

The emission wavelength is usually given as a single number, which is the wavelength of peak emission. However, it is often useful to consider the emission wavelength as a range of wavelengths rather than a single number. The width of the emission wavelength range depends on the type of light source. For example, an incandescent light bulb will have a very broad emission wavelength range, while a laser will have a very narrow emission wavelength range.

The emission wavelength can also be used to determine the power of the light source. The power of a light source is measured in watts, which is determined by the amount of light emitted by the source. The wavelength of the light emissions also plays a role in the power of the light.

2.1.2 Light coherence

The coherence of a light source is a measure of the degree to which its emissions are temporally correlated. This is an important consideration because it can affect the visibility of interference patterns. In general, it is desirable for a light source to have a high degree of coherence, as this can improve the quality of the light. Coherence can be measured in terms of the temporal correlations between the light's amplitude, phase, or frequency. A light source with a high degree of coherence will produce light waves with the same characteristics over time, while a light source with a low degree of coherence will produce light waves with varying characteristics.

The term 'degree of coherence' is often used in the same context as the term 'spatial coherence.' While the two terms are not identical, they both describe how temporally or spatially consistent the characteristics of a single light wave are. There are many types of systems that require very high spatial or temporal coherence, including some that rely on the production of interference patterns.

We can differentiate two types of coherence: temporal and spatial. What sets temporal and spatial coherence apart is that temporal coherence denotes the correlation between waves observed at separate times, while spatial coherence represents the correlation between waves at different places—lateral or longitudinal. Temporal coherence is related to the consistency of waveforms over time, and it is often used in signal processing for signal analysis. Spatial coherence is often used to measure the degree to which two points are correlated in space. This type of coherence is important in understanding wave propagation and interference patterns, especially in optics and acoustics.

2.1.3 Emission power

The emission power of a light source is the power of the light it emits. This is an important consideration because it can affect the brightness of the light. The emission power of a light source is usually measured in lumens, which is a measure of the total amount of visible light emitted from the source. The higher the lumen rating, the brighter the light. Different types of light sources have different emission powers, and some are more efficient than others. For example, LED lights generally have a higher lumen output than traditional incandescent bulbs. Not only is the emission power of a light source important for determining the brightness of the light, but it can also affect the quality of the light. For example, a light source with a higher emission power can produce a more intense light, which can be useful for certain applications. On the other hand, a light source with a lower emission power may be more suitable for creating a softer, more diffused light. The emission power of a light source is just one of the many factors to consider when choosing the right light for your needs.

2.1.4 Light polarization

Light is an electromagnetic wave, that is composed of an electric and magnetic field that are always perpendicular to each other. The 'polarization' of this wave describes the orientation of the electric field. In linear polarization, the fields vibrate in one direction. For circular or elliptical polarization, the fields rotate at a constant rate in the same plane as the wave advances.

There are several different types of light polarization, including:

1. **Linear polarization:** linear polarization occurs when the electric field of the light wave oscillates in a single plane, such as vertically or horizontally.
2. **Circular polarization:** circular polarization occurs when the electric field of the light wave rotates in a circular direction around the direction of propagation. There are two types of circular polarization: right-handed and left-handed.
3. **Elliptical polarization:** elliptical polarization occurs when the electric field of the light wave oscillates in two perpendicular planes, causing the polarization ellipse to rotate as the wave propagates. The degree of ellipticity is defined by the ratio of the minor axis to the major axis of the polarization ellipse.
4. **Random polarization:** random polarization, also known as unpolarized light, occurs when the electric field of the light wave oscillates in random directions.
5. **Hybrid polarization:** hybrid polarization occurs when the electric field of the light wave has multiple components, such as linear and circular polarization.

Certain materials have the ability to transmit incident unpolarized light in only one direction. A polarizing sheet has an orientation that allows it to transmit only the electric field vector parallel to its axis, while completely absorbing the perpendicular component. This means that light passing through the sheet becomes linearly polarized in the direction of the transmission axis. It is also worth noting that polarizers reduce the intensity of the incident light beam to some extent.

A retarder (or wave plate) takes an incoming light wave and splits it into two orthogonal linear polarization components. Depending on the phase difference, the light that comes out could have a different polarization from what it originally had. It's important to remember that a retarder won't polarize unpolarized light and ideally won't reduce the intensity of the beam. Quarter-wave and half-wave plates are two of the most commonly used retarders. A quarter-wave plate can be used to switch between linear and circular polarization, whereas a half-wave plate is an effective polarization rotator, capable of changing the angle of linear polarization by twice the angle between its optic axis and the initial direction of polarization.

2.2 Incandescent sources

Incandescent lamps produce thermal radiation from an electrically heated filament. The filaments are usually tungsten, allowing the lamp to operate at high temperatures. The filament's temperature is generally between 2000 and 3000 K, resulting in a warm, orange-yellow light. As an electric current heats the filament, it emits thermal radiation from visible light and some infrared radiation. The amount of radiation emitted depends on the surface temperature of the filament, which is determined by the amount of current applied to it.

Fluorescent lamps produce light by passing an electric current through a gas-filled tube that contains mercury vapor and other inert gases. The electric current excites the mercury atoms, causing them to emit ultraviolet radiation. This ultraviolet radiation is converted into visible light when it strikes the phosphor coating inside the lamp.

Fluorescent lamps are much more energy-efficient than traditional incandescent bulbs since they produce more light for less power input. Fluorescent lamps have a longer lifespan than incandescent bulbs, lasting up to 10 times as long. Fluorescent lamps also produce far less heat, making them better suited for many applications.

Some advantages of incandescent lamps are:

1. They produce a wide range of light, including all the visible spectrum colors.
2. They can be dimmed relatively easily.
3. They are inexpensive to produce.

The disadvantages of incandescent lamps include the following:

1. They are relatively inefficient, only converting about 10% of the electrical energy they consume into visible light.
2. They have a relatively short lifespan, typically around 1000 h.
3. They are sensitive to vibration, which can further shorten their lifespan.

Overall, incandescent lamps are a good choice for general lighting applications where cost is a significant consideration. However, their inefficiency makes them less ideal for applications where energy conservation is a priority.

Incandescent sources emit a continuous spectrum of light that can be described using Planck's law of blackbody radiation. This spectrum is composed of a range of wavelengths of visible light and infrared and ultraviolet radiation. The peak intensity of the light produced by an incandescent source depends on its temperature, with hotter sources emitting more energy in the higher wavelength (blue) of the spectrum. Planck's equation describes how an incandescent source emits light (2.1).

$$B_\nu(T) = \frac{2\nu^2}{c^2} \frac{h\nu}{e^{h\nu/kT} - 1}, \tag{2.1}$$

where: $B_\nu(T)$ is the spectral radiance, h is the Planck constant, c is the speed of light in a vacuum, k is the Boltzmann constant, ν is the frequency of the electromagnetic radiation, and T is the absolute temperature of the body.

In other words, an incandescent source emits a continuous spectrum of light, with different frequencies of light produced at different temperatures. This means that a single incandescent source can have a range of different colors. The light spectrum emitted by an incandescent source is also known as 'blackbody radiation' because it resembles the curve of light produced when a black object is heated up. Planck's law is a valuable tool for describing the behavior of incandescent sources, and it can help us understand why they produce the colors they do.

2.3 Light emitting diodes

Whereas an incandescent source emits a broad output spectrum of light, a light emitting diode, or LED, can generate light close to certain wavelengths. This provides the opportunity to create tailored light sources for different applications.

LEDs can be used to produce specific colors of light, making them ideal for applications that require more precise lighting than what is available from an incandescent source. For example, they are often used in medical applications such as endoscopy and dental equipment where exact color of light is necessary. They are also commonly used in decorative lighting and displays due to their ability to produce vibrant colors and patterns. Finally, LEDs are highly efficient, using less energy than traditional light sources, making them an environmentally friendly choice. LEDs were first reported in 1927 by Russian scientist Oleg Losev [9].

LEDs are a type of solid-state semiconductor device that emits light when a current passes through it. They are formed by a junction between a p-type and an n-type semiconductor. A current flows only from the p-side (anode) to the n-side (cathode). LEDs emit a particular wavelength of light depending on the semiconductor materials used in their construction. When the current flows from the p-side element (anode) to the n-type elements (cathode) electrons and holes recombine at the junction. It is possible then for an electron to recombine with a hole emitting a photon in the process. The energy of the photon will approximately be the energy of the bandgap between the two semiconductors, and is given by the equation (2.2):

$$E_\mathrm{p} = \frac{hc}{\lambda} = \frac{1.25\ \mathrm{eV} \cdot \mu\mathrm{m}}{\lambda} \tag{2.2}$$

where: h is the Plack's constant, c is light's speed in m s^{-1}, and λ is the source wavelength in μm.

A single material cannot cover the entire spectrum of light, meaning that each unique combination of semiconductors produces a unique energy bandgap. By combining semiconductor materials it is possible to modify their band gaps, and create an LED that emits any desired wavelength.

For example, galium arsenide (GaAs) has a bandagap of 1.43 eV (λ = 860 nm); adding aluminum to the compound to create aluminum aalium arsenide ($Al_xGa_{1-x}As$) allows one to modify the bandgap from 1.5 eV to 2.1 eV (590 nm < λ < 840 nm). Indium gallium nitride ($In_xGa_{1-x}N$) allows for compounds with a bandgap of 0.67 eV and 3.3 eV (390 nm < λ < 1750 nm). Blue emission can be formed using gallium nitride (GaN), or zinc zelenide (ZnSe). Table 2.1 shows some common LEDs compounds and their respective emission wavelength [1].

LEDs generate photons through spontaneous emission; this occurs when electrons, already in an elevated energy state, move back to their original energy level independently of other electrons making the same transition. This is one of the reason for LEDs' relative wide spectral linewidth (around 10–50 nm).

The are several advantages of LEDs over incandescent light sources. For example:

1. Low power: typical LEDs have an operating voltage of less than 5 V and a current of 20 mA.
2. Long life: LEDs, when properly operated, have a useful life of thousands of hours.
3. Cool light: there is low emission of heat when operating an LED, most of the applied power is converted into photons. So, LEDs are cold to the touch.
4. Wavelength output: as mentioned before, we can engineer the bandgap of the semiconductors to emit light with a specific wavelength.

Among the disadvantages of LEDs are:

1. Fabrication technology: growing the necessary semiconductors with the required purity is not a trivial matter. Some compounds like GaN were not commercially available until 1989 due to their difficulty of fabrication.
2. Low coherence: due to their spontaneous emission LEDs have low coherence when compared to Laser sources. So they have limited use in interferometers.

Table 2.1. LEDs compounds at their emission wavelength.

Compound	Wavelength (nm)
Aluminum gallium nitride (AlGaNi)	⩽400
Silicon carbide (SiC)	450–500
Gallium phosphide (GaP)	500–570
Aluminum indium gallium phosphide (AlInGaP)	570–620
Gallium arsenide (AlGaAs)	⩾760

2.4 Laser

The invention of the laser was a result of several decades of research in the field of optics, specifically in the development of masers (Microwave Amplification by Stimulated Emission of Radiation). In 1917, Albert Einstein first proposed the theory of stimulated emission, which describes the process by which atoms can be stimulated to emit light in a specific direction [4, 16].

In the 1950s, Charles Townes, Arthur Schawlow, and their colleagues were working on developing masers, which used microwaves to create coherent radiation. They realized that they could extend their work to the visible spectrum and create a device that would amplify and focus light in a similar way to the maser. Townes and Schawlow published a paper in 1958 outlining the theoretical principles for such a device, which they called an ‘optical maser’ [18].

Theodore H Maiman, a physicist at Hughes Research Laboratories, was the first person to actually build a working optical maser [10]. He used a synthetic ruby crystal as the lasing medium, which was excited by a high-intensity flash lamp to emit a narrow beam of red light. On May 16, 1960, Maiman successfully demonstrated his device, which he called a ‘ruby laser.’

Charles H Townes and Nikolay Gennadiyevich Basov, along with Alexander M Prokhorov, shared the 1964 Nobel Prize in Physics ‘for fundamental work in the field of quantum electronics, which has led to the construction of oscillators and amplifiers based on the maser-laser principle’ [3]. Townes and Basov were recognized for their work on the theoretical basis of the laser, while Prokhorov was recognized for his contributions to the development of the maser.

The basic principle behind lasers is the emission of light through stimulated emission. When atoms or molecules are excited, they release energy in the form of photons, which can be thought of as packets of light energy. In a laser, this process is amplified and directed into a beam of light of a single wavelength and high coherence [4, 11, 13]. The result is a powerful and precise light source that has many practical applications.

One of the key advantages of lasers is their ability to produce light with a very narrow bandwidth and high spectral purity. This makes them ideal for optical sensors, which measure a wide range of physical parameters such as temperature, pressure, and strain. Optical sensors can be used in various fields, from manufacturing to environmental monitoring, and are critical for the development of smart cities and intelligent transportation systems.

One example of an optical sensor that uses lasers is the fiber-optic temperature sensor, which is used to measure temperature in harsh environments where traditional sensors cannot be used. Fiber-optic sensors work by sending a beam of light through a fiber-optic cable and measuring the change in the refractive index of the fiber as it is exposed to temperature changes. The resulting change in the wavelength of the light can then be used to calculate the temperature.

Another application of lasers in optical sensors is in the development of chemical sensors, which are used to detect the presence of gases and other substances in the environment [4, 6, 14]. Chemical sensors work by measuring the absorption or scattering of light by the target molecule, and lasers are used to provide a high-intensity light source that can be tuned to the specific absorption or scattering spectrum of the target molecule. This makes them highly sensitive and specific, allowing for the detection of even trace amounts of the target substance.

Laser-based sensors are also used in structural health monitoring, which is the process of monitoring the health of a structure such as a building or bridge to detect any signs of damage or deterioration. This is important for ensuring the safety of structures and preventing catastrophic failures. Laser-based sensors can be used to measure parameters such as strain, vibration, and deformation, providing a highly accurate and sensitive method of detecting any changes in the structure [6, 14].

The use of lasers in optical sensors has also revolutionized the field of biomedical imaging. Optical coherence tomography (OCT), for example, uses lasers to produce high-resolution images of biological tissues such as the eye and skin. OCT works by measuring the reflections of light from within the tissue, allowing for the creation of detailed 3D images of the structure of the tissue [2]. This technique has many applications in the diagnosis and treatment of diseases such as macular degeneration and skin cancer.

In addition to their applications in optical sensors, lasers have many other practical uses. They are widely used in the manufacturing industry for cutting, drilling, and welding, and are also used in the entertainment industry for light shows and laser displays. Lasers are also used in military applications, such as target designation and range finding.

A laser typically consists of the following components:

1. **Gain medium:** this is the material that amplifies the light waves through stimulated emission of photons. It can be a solid, liquid, gas, or semi-conductor depending on the type of laser (figure 2.2).
2. **Pump source:** the gain medium is excited using a pump source such as a flash lamp, diode, or another laser beam, which supplies energy to the gain medium to bring it to the desired energy level.
3. **Optical resonator:** the optical resonator consists of two mirrors placed at either end of the gain medium. One of the mirrors is partially transparent, allowing some of the laser light to escape as the laser beam.
4. **Laser beam:** the laser beam is the final output of the laser, consisting of coherent light waves with a narrow bandwidth and high intensity.

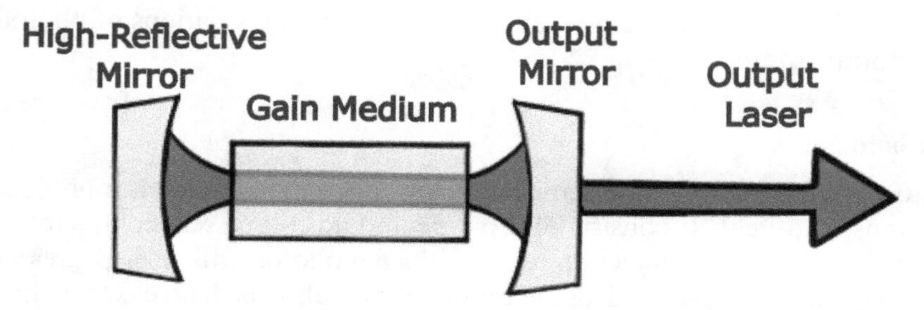

Figure 2.2. Typical components of a laser. Image created by L Argueta-Slaughter.

5. **Control and feedback mechanisms:** lasers often require sophisticated control and feedback mechanisms to maintain the precise conditions necessary for optimal performance, such as temperature control, beam alignment, and stabilization of the optical resonator.

2.4.1 Safety classes

Laser safety classes are a standardized system used to classify lasers based on their potential for causing eye or skin damage. The classification system is based on the maximum power output of the laser and the wavelength of the emitted light. There are currently six laser safety classes, designated as Class 1, 1M, 2, 2M, 3R, 3B, and 4, with Class 1 being the safest and Class 4 being the most hazardous. Here's a brief description of each class:

Class 1: these lasers are considered safe and do not pose any hazard to eyes or skin.
Class 1M: these lasers are similar to Class 1 but may be hazardous if viewed directly with optical instruments.
Class 2: these lasers emit visible light and are considered safe for momentary exposure (less than 0.25 s) as the blink reflex protects the eyes.
Class 2M: these lasers are similar to Class 2 but may be hazardous if viewed directly with optical instruments.
Class 3R: these lasers are low-power lasers that may pose a hazard if viewed directly with the naked eye, but are not considered hazardous when viewed indirectly.
Class 3B: these lasers are moderate-power lasers that can cause eye and skin damage if viewed directly, and pose a fire hazard.
Class 4: these lasers are high-power lasers that can cause severe eye and skin damage, and may also pose a fire hazard and create other safety hazards due to their power output. Special safety precautions are required when operating Class 4 lasers.

2.5 Photodiodes, and phototransistors

Photodiodes and phototransistors are two types of semiconductor devices that convert light energy into electrical energy. They are widely used in various applications, such as optical communication, sensing, and imaging. This section will discuss the working principles, characteristics, and applications of photodiodes and phototransistors.

2.5.1 Photodiodes

A photodiode is a p–n junction semiconductor device that generates a photocurrent when exposed to light. It consists of a p-type and an n-type semiconductor region, which are separated by a depletion region. When a photon with energy greater than the bandgap energy of the semiconductor material is absorbed, an electron–hole pair is created. The electron is then swept into the n-type region, and the hole is swept into the p-type region due to the built-in electric field in the depletion region. This creates a current flow in the external circuit, which is the photocurrent [1, 17].

The photocurrent generated by a photodiode is proportional to the incident light power, and it increases with increasing light intensity. However, it is also affected by other factors, such as the wavelength of the incident light, the temperature, and the bias voltage applied to the photodiode. The responsivity of a photodiode is defined as the ratio of the photocurrent to the incident optical power, and it is usually specified in units of A/W.

The dark current of a photodiode is the current that flows through the device in the absence of light. It is mainly due to the thermal generation of electron–hole pairs and the diffusion of carriers across the depletion region. The dark current increases with increasing temperature and bias voltage, and it can limit the sensitivity of the photodiode. Therefore, it is important to minimize the dark current in photodiode design.

Photodiodes can be classified into two types: p–n junction photodiodes and PIN photodiodes. The p–n junction photodiode has a typical response time of a few nanoseconds and a quantum efficiency of up to 80% at the peak wavelength. It is commonly used in low-speed applications, such as optical sensing and control. The PIN photodiode has a wider depletion region, which reduces the capacitance and increases the response time. It has a higher quantum efficiency and can operate at higher speeds, making it suitable for high-speed optical communication applications [13, 15, 17].

2.5.2 Phototransistors

A phototransistor is a bipolar transistor that has a light-sensitive base region. It operates in a similar way to a photodiode, but it has an additional amplification stage. When photons are absorbed by the base region, they create electron–hole pairs that increase the base current of the transistor [7, 13]. This, in turn, increases the collector current, which is the output current of the device. Therefore, a phototransistor can provide a larger output current than a photodiode for the same incident light power.

The amplification factor of a phototransistor, also known as the current gain, is the ratio of the collector current to the base current. It can be as high as 1000 for some phototransistor designs [1, 15]. The frequency response of a phototransistor is limited by its transit time, which is the time it takes for carriers to travel across the base region. It can range from a few nanoseconds to several microseconds, depending on the device geometry and bias conditions.

Phototransistors can be classified into two types: bipolar phototransistors and field-effect phototransistors. The bipolar phototransistor has a similar structure as a standard bipolar transistor, with a light-sensitive base region. It has a high gain and a fast response time but also a high capacitance, which limits its high-frequency performance. On the other hand, the field-effect phototransistor has a light-sensitive gate region instead of a base region. It operates based on the modulation of the gate-source current by the incident light. It has a lower capacitance and a higher speed than the bipolar phototransistor but also a lower gain.

2.5.2.1 Applications

Photodiodes and phototransistors have numerous applications in various fields. One of the most common applications is in optical communication systems, such as fiber-optic communication and free-space optical communication. Photodiodes and phototransistors are used as receivers to convert the optical signals into electrical signals. The high speed and sensitivity of photodiodes and phototransistors make them suitable for high-speed communication systems.

Photodiodes are also used in sensing applications, such as optical sensors and detectors. They can be used to detect the presence, intensity, and wavelength of light. For example, photodiodes can be used in optical smoke detectors to detect the presence of smoke particles by measuring the scattered light. They can also be used in imaging applications, such as digital cameras and video cameras. Photodiode arrays can be used to capture and process images with high resolution and sensitivity.

Phototransistors are used in applications where a larger output current is required, such as in optical switches and relays. They can also be used in optical sensing applications where the signal needs to be amplified. For example, phototransistors can be used in photoelectric sensors to detect the presence of objects and to control industrial processes.

Photodiodes and phototransistors are important semiconductor devices that convert light energy into electrical energy. They have different working principles, characteristics, and applications, which make them suitable for various applications in optical communication, sensing, and imaging. The design and optimization of photodiodes and phototransistors are essential for achieving high sensitivity, speed, and reliability. Therefore, the development of new materials and fabrication techniques for photodiodes and phototransistors is an active research area in semiconductor technology.

2.6 Image sensors: CCD, and CMOS

In modern scientific research, the use of cameras has become an essential tool to capture high-quality images for analysis and data collection. Two types of cameras commonly used in science experiments are charge-coupled device (CCD) and complementary metal-oxide–semiconductor (CMOS) cameras. These cameras differ in their construction and operation, and each has its advantages and disadvantages.

2.6.1 Charge-coupled device (CCD) camera

CCD cameras were first developed in the late 1960s and are now commonly used in scientific research due to their high sensitivity and low noise. A CCD camera consists of a silicon-based chip that has an array of light-sensitive pixels, which convert light into electrical signals that can be stored and processed. When light strikes the pixels, it generates an electrical charge proportional to the intensity of the light. The charge is then transferred from pixel to pixel until it reaches the output amplifier. The amplifier converts the charge into an analog voltage that is then digitized and stored [5, 19].

CCD cameras have several advantages that make them suitable for scientific experiments. Firstly, they have a high sensitivity to light, which makes them useful in low-light conditions. Secondly, they have low noise levels, which means that they produce high-quality images with minimal distortion. Thirdly, they have a wide dynamic range, which means that they can capture both bright and dark regions in the same image. Lastly, they are capable of producing high-resolution images, which is essential in scientific research.

However, CCD cameras also have some disadvantages that must be considered. Firstly, they are relatively expensive compared to CMOS cameras. Secondly, they have a slower readout time, which limits their use in high-speed applications. Lastly, they consume more power than CMOS cameras, which can be a problem in battery-operated systems.

2.6.2 Complementary metal-oxide–semiconductor (CMOS) camera

CMOS cameras were first introduced in the early 1990s and have since become increasingly popular due to their lower cost and higher speed compared to CCD cameras [8, 20]. A CMOS camera also consists of a silicon-based chip with an array of light-sensitive pixels. However, in contrast to CCD cameras, each pixel in a CMOS camera has its amplifier, which converts the charge into a voltage. This reduces the readout time significantly, making CMOS cameras more suitable for high-speed applications.

CMOS cameras also have some advantages over CCD cameras [5, 19]. Firstly, they are cheaper to manufacture, making them more accessible to researchers with limited budgets. Secondly, they consume less power than CCD cameras, making them ideal for battery-operated systems. Lastly, they have a higher readout speed than CCD cameras, which makes them more suitable for high-speed applications.

However, CMOS cameras also have some disadvantages. Firstly, they have a lower sensitivity to light compared to CCD cameras, making them less suitable for low-light conditions. Secondly, they have higher noise levels, which can result in lower image quality. Lastly, they have a smaller dynamic range than CCD cameras, which means they cannot capture both bright and dark regions in the same image.

2.6.3 Comparison

Both CCD and CMOS cameras have various applications in scientific experiments. CCD cameras are commonly used in astronomy, where they are used to capture high-resolution images of stars and galaxies. They are also used in biology, where they are used to capture images of cells and tissues for analysis. In addition, CCD cameras are used in microscopy, where they are used to capture images of specimens for analysis.

On the other hand, CMOS cameras are commonly used in high-speed imaging applications, such as particle tracking and fluid dynamics.

Table 2.2. Comparison between CCD and CMOS cameras.

Feature	CCD cameras	CMOS cameras
Sensitivity	High	Moderate
Noise	Low	Moderate
Dynamic range	High	Low
Power consumption	High	Low
Cost	High	Low
Frame rate	Moderate	High
Resolution	High	Moderate

When choosing between CCD and CMOS cameras for a science experiment, there are several factors to consider. Table 2.2 summarizes the main differences between these two types of cameras [5, 20].

As shown in the table, CCD cameras have higher sensitivity, lower noise, and better dynamic range than CMOS cameras. However, CMOS cameras have lower power consumption, are less expensive, and have a higher frame rate than CCD cameras. The choice between CCD and CMOS cameras ultimately depends on the specific needs of the experiment.

References

[1] Bhattacharya P 1997 *Semiconductor Optoelectronic Devices* 2nd edn (Englewood Cliffs, NJ: Prentice-Hall)

[2] Drexler W, Fujimoto J G and Swanson E A 2015 *Optical Coherence Tomography: Technology and Applications* (Berlin: Springer)

[3] The Nobel Foundation 1964 *The Nobel Prize in Physics 1964* https://www.nobelprize.org/prizes/physics/1964/summary/ (accessed 13 April 2023)

[4] Hecht J 1991 *City of Light: The Story of Fiber Optics* (Oxford: Oxford University Press)

[5] Kalra R and Kaur H 2017 A review on CMOS and CCD image sensors *Int. J. Innov. Res. Sci. Eng. Technol.* 6 25–31

[6] Hung O-n, *et al* 2020 Application of laser technology *Sustainable Technologies for Fashion and Textiles* (Cambridge: Woodhead Publishing) pp 163–87

[7] Cuevas A and Balbuena M A 1989 Review of analytical models for the study of highly doped regions of silicon devices *IEEE Trans. Electron Devices* **36** 553–60

[8] Liao Y-J and Chen C-C 2015 High-speed CMOS image sensor design for image acquisition systems *J. Electron. Sci. Technol.* **13** 241–6

[9] Losev O V 1928 Luminous carborundum detector and detection effect and oscillations with crystals *Philos. Mag.* **5** 1024–44

[10] Maiman T H 1960 Stimulated optical radiation in ruby *Nature* **187** 493–4

[11] Pedrotti F L, Pedrotti L M and Pedrotti L S 2017 *Introduction to Optics* (Cambridge: Cambridge University Press)

[12] Rohde R A 2007 *Solar Spectrum* https://commons.wikimedia.org/wiki/File:Solar_Spectrum.png

[13] Saleh B E A and Carl Teich M 2019 *Fundamentals of Photonics* (New York: Wiley)

[14] Santos J L, Arregui F J and Matias I R 2010 *Optical Fiber Sensors: A Practical Approach* (New York: Springer)
[15] Shah J M 2001 *Photodiodes Encyclopedia of Materials: Science and Technology* (Amsterdam: Elsevier) pp 7016–21
[16] Siegman A E 1986 *Lasers* (Mill Valley, CA: University Science Books)
[17] Sze S and Ng K 2006 *Physics of Semiconductor Devices* 3rd edn (New York: Wiley)
[18] Townes C H and Schawlow A L 1958 Infrared and optical masers *Phys. Rev.* **112** 1940–9
[19] Van Bommel T and Chi N 2016 Choosing the right camera for your microscopy application *Microsc. Today* **24** 38–43
[20] Wong K C 2009 Performance Comparison between CMOS and CCD sensors for industrial inspection applications *IEEE Trans. Instrum. Meas.* **58** 4238–43

IOP Publishing

Optical Sensors
An introduction with lab demonstrations
Victor Argueta-Diaz

Chapter 3

Maxwell equations

3.1 Introduction

Classic electromagnetic theory was formulated by James Clerk Maxwell (13 June 1831–5 November 1879) in 1865. Maxwell worked on the models Gauss, Faraday, and Ampère developed and found the connections between electric and magnetic fields. These connections can be represented in what are known as Maxwell's equations.

Maxwell's equations can be described as follows:

1. Gauss's law for electric fields helps us describe the electric field calculated by a charge distribution.
2. Gauss's law for magnetic fields explains that the net magnetic flux through a closed surface is always zero.
3. Faraday's law, which says a time-varying magnetic field produces an electric field.
4. Ampère–Maxwell's law which says a time-varying electric field produces a magnetic field.

Equations (3) and (4) imply that when a time-varying magnetic field creates an electric field, this said electric field will develop another magnetic field. As the cycle continues, an electromagnetic wave will be created that will propagate through space.

Maxwell's equations can be expressed in either an integral or differential notation. The following section will explain these equations and their interpretation in both notations. These notations are shown in table 3.1.

3.2 Gauss's law for electric fields

We will first start with Gauss's law for electric fields in their integral form.

doi:10.1088/978-0-7503-4876-8ch3

Table 3.1. Maxwell's equations in integral and differential form.

Equation name	Integral form	Differential form
Gauss' law of electric fields	$\oint_S (\mathbf{E} \cdot \hat{n})\, ds = \frac{q_{enc}}{\varepsilon_0}$	$\nabla \cdot \mathbf{E} = \frac{\rho}{\varepsilon_0}$
Gauss' law of magnetic fields	$\oint_S (\mathbf{B} \cdot \hat{n})\, ds = 0$	$\nabla \cdot \mathbf{B} = 0$
Faraday's law	$\oint_S \mathbf{E} \cdot d\vec{l} = -\frac{d}{dt} \int_S (\mathbf{B} \cdot \hat{n})\, ds$	$\nabla \times \mathbf{E} = -\frac{\partial \mathbf{B}}{\partial t}$
Ampère–Maxwell equation	$\oint_S \mathbf{B} \cdot d\vec{l} = \mu_0 \left[I_{enc} + \varepsilon_0 \frac{d}{dt} \int_S (\mathbf{E} \cdot \hat{n})\, ds \right]$	$\nabla \times \mathbf{B} = \mu_0 \left(\vec{J} + \varepsilon_0 \frac{\partial \mathbf{E}}{\partial t} \right)$

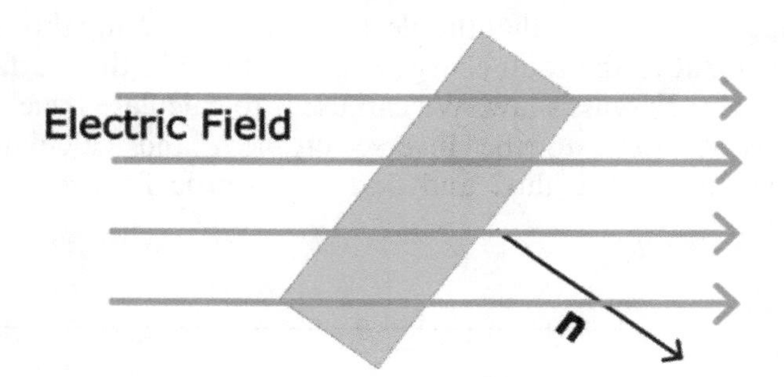

Figure 3.1. Electric flux through a surface.

$$\oint_S (\mathbf{E} \cdot \hat{n})\, ds = \frac{q_{enc}}{\varepsilon_0} \tag{3.1}$$

where: $\mathbf{E}$ is the electric field vector, $\hat{n}$ is the unit vector normal to the surface, q_{enc}, is the electric charge contained inside the closed surface, and ε_0 is the electric permittivity of free space with an approximate value of $8.854\ 18 \times 10^{-12}$ C (Vm)$^{-1}$.

The left side of the equation represents the electric flux through a closed surface, while the right side represents the net charge divided by the electric permittivity of the material inside that closed surface. The electric flux is the amount of the electric field that goes through a surface, as shown in figure 3.1. In Gauss's law, we are interested in the perpendicular component of the electric field with respect to the surface, whence the use of a dot product between the electric field and a normal unit vector to the surface $\hat{n}$.

When considering the electric flux through a closed surface, we need to consider the net flux, that is, the total number of electric field lines that go into the surface as well as come out of the surface. For example, in figure 3.2a we can see that the same number of electric field lines going into the closed surface as coming out of it, so the net flux will be zero. This happens when the electric field's source (and end) is outside the close surface. On the other hand, if we place a net charge inside the close surface, then we will have a non-zero electric flux. If the net charge is positive, as in figure 3.2b, we will have a positive flux, or if it's a negative net charge, as in figure 3.2c, we will have a negative flux.

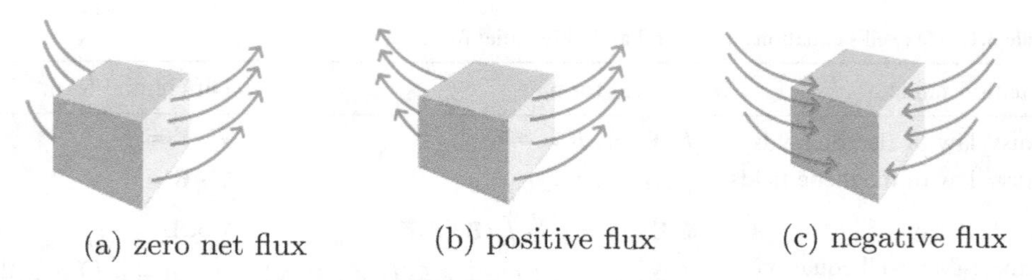

Figure 3.2. Electric flux through different surfaces.

Therefore, Gauss's law states that the electric field flux passing through any closed surface is proportional to the total charge contained within that surface. There are several applications for Gauss's law. We can use it to calculate the electric flux of a given charge distribution, do the inverse problem and calculate the charge distribution given an electric flux, and find the electric field caused by a given charge distribution.

Example 3.1. Using Gauss's law, find the electric field created by a point charge.

To solve this problem, we can create a spherical Gaussian surface of radius r, with our charge at its center, as shown in figure 3.3.

Working with a sphere as our Gaussian surface allows us to do some simplification to Gauss's law equation. For example, at all points on our spherical surface, the electric field will point in the same direction as our normal vector to the surface, Also, the electric field will be constant at all points along the surface. So we can rewrite equation (3.1) as follows:

$$\begin{aligned} \oint_S (\mathbf{E} \cdot \hat{n})\, ds &= \frac{q_{\text{enc}}}{\varepsilon_0} \\ \oint_S E\, ds &= \frac{q_{\text{enc}}}{\varepsilon_0} \\ E \oint_S ds &= \frac{q_{\text{enc}}}{\varepsilon_0} \\ E 4\pi r^2 &= \frac{q_{\text{enc}}}{\varepsilon_0} \\ E &= \frac{1}{\varepsilon_0} \frac{q_{\text{enc}}}{4\pi r^2} \hat{r} \end{aligned} \tag{3.2}$$

which corresponds to the electric field calculated by Coulomb's law. We added the $\hat{r}$ vector to indicate the radial direction of the electric field.

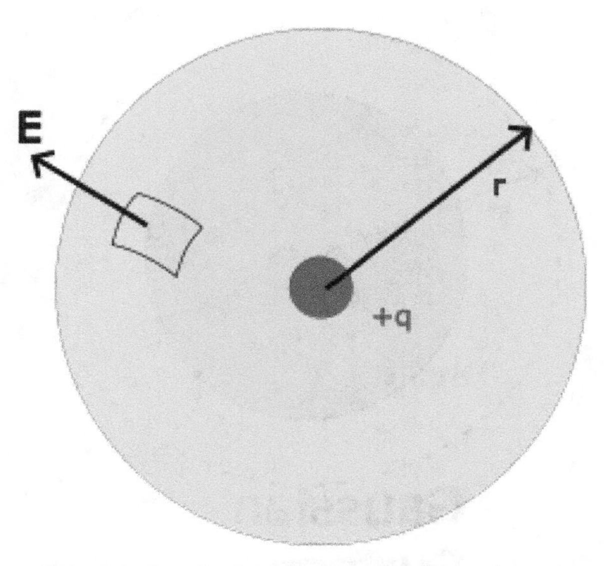

Figure 3.3. Gaussian sphere surrounding a point charge q.

Example 3.2. Calculate the electric field of a uniform charged dielectric sphere. Plot a graph of E versus r.

Imagine a sphere of radius a with a uniform charge ρ_0. To calculate the electric field everywhere in space, we will need to calculate the electric field inside the sphere, $r \leqslant a$, and outside the sphere, $r \geqslant a$. We will use a spherical Gaussian surface as shown in figure 3.4.

We will first calculate the charge enclosed by our Gaussian surface of radius r.

$$\begin{aligned} q_{\text{enc}} &= \int_v \rho_0 dv \\ &= \rho_0 \int_v dv \\ &= \rho_0 \int_{\phi=0}^{2\pi} \int_{\theta=0}^{\pi} \int_{r=0}^{r} r'^2 \sin\theta dr d\theta d\phi \\ q_{\text{enc}} &= \rho \frac{4}{3} \pi r^3 \end{aligned} \tag{3.3}$$

we introduce r' to distinguish our variable from our limits of integration. Here the volume integral represents the volume of our Gaussian surface

The flux can then be calculated as

$$\begin{aligned} \oint_S (\mathbf{E} \cdot \hat{n}) ds &= E \oint_S ds \\ &= E_r \int_{\phi=0}^{2\pi} \int_{\theta=0}^{\pi} r^2 \sin\theta d\theta d\phi \\ &= E_r 4\pi r^2 \end{aligned} \tag{3.4}$$

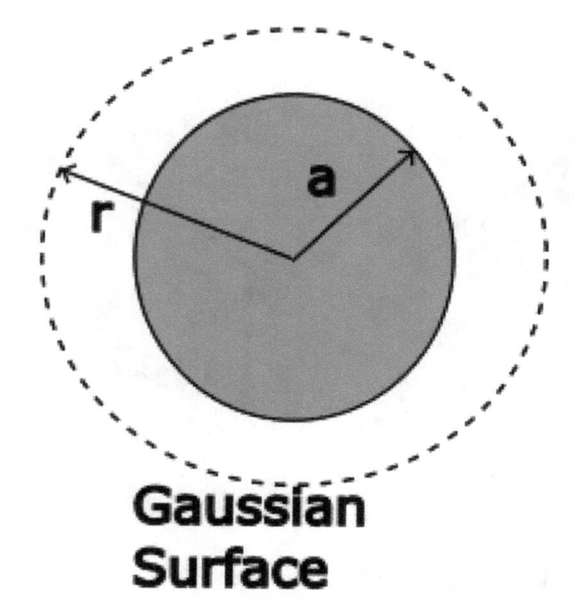

Figure 3.4. Gaussian surface for a uniformly charged sphere or radius a.

Equating equation (3.4) to (3.3) we get:

$$\begin{aligned} E_r 4\pi r^2 &= \rho_0 \frac{4}{3}\pi r^3 \\ E_r &= \frac{\rho_0}{3} r \; 0 < r \leqslant a \end{aligned} \tag{3.5}$$

We now proceed to calculate the electric field outside the charged sphere. Because our spherical Gaussian surface is larger than the charged sphere, the enclosed charge will be the net charge of the sphere. Notice how in equation (3.6) the limits of integration change for our first integral.

$$\begin{aligned} q_{\text{enc}} &= \int_v \rho_0 dv \\ &= \rho_0 \int_v dv \\ &= \rho_0 \int_{\phi=0}^{2\pi} \int_{\theta=0}^{\pi} \int_{r=0}^{a} r'^2 \sin\theta dr d\theta d\phi \\ q_{\text{enc}} &= \rho \frac{4}{3}\pi a^3 \end{aligned} \tag{3.6}$$

When calculating the flux, we obtain the same result as in equation (3.4), So the electric field outside the charged sphere is:

$$\begin{aligned} E_r 4\pi r^2 &= \rho_0 \frac{4}{3}\pi a^3 \\ E_r &= \frac{\rho_0 a^3}{3r^2} \; r \geqslant a \end{aligned} \tag{3.7}$$

Our results can be summarized as follows:

$$E_r = \begin{cases} \frac{\rho_0}{3} r, & 0 < r \leqslant a \\ \frac{\rho_0 a^3}{3r^2}, & r \geqslant a \end{cases} \tag{3.8}$$

From (3.5) we can see that the electric field increases linearly inside the sphere, to reach a maximum value at the edge of the sphere. Once we are outside the sphere, the electric field has a dependency on $\approx 1/r^2$, which makes it decrease rapidly, as shown in figure 3.5.

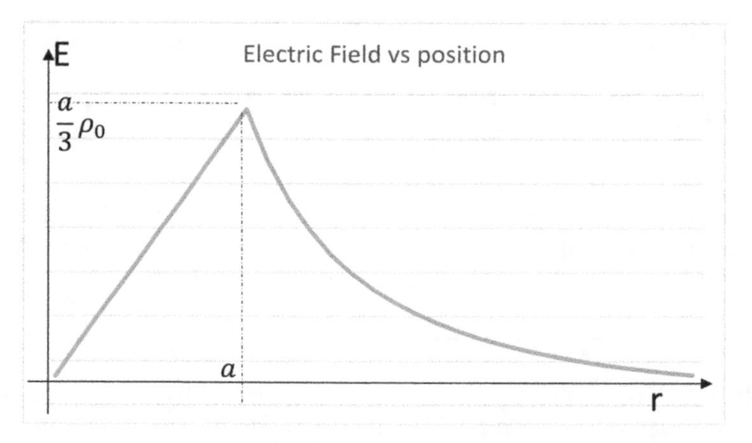

Figure 3.5. Electric field versus r for a uniform charged sphere of radius a.

Gauss's law for electric fields—differential form

In its differential form Gauss's law can be expressed as:

$$\nabla \cdot \mathbf{E} = \frac{\rho}{\varepsilon_0} \tag{3.9}$$

where: ∇ is the gradient operator and is defined, in Cartesian coordinates, as:

$$\nabla = \hat{\imath}\frac{\partial}{\partial x} + \hat{\jmath}\frac{\partial}{\partial y} + \hat{k}\frac{\partial}{\partial z} \tag{3.10}$$

where $\hat{\imath}$, $\hat{\jmath}$, and $\hat{k}$ are the unit vectors in the x, y, and z directions, respectively.

The left side of the equation represents the divergence of the electric field. That is, how is the electric field moving, either away or towards, a specific point in space. The right side of the equation represents the electric charge density, in C m^{-3}, divided by the electric permittivity, ε_0. Gauss's law in its differential form shows us that the electric field diverges from a positive charge, and converges from a negative charge, as shown in figure 3.6.

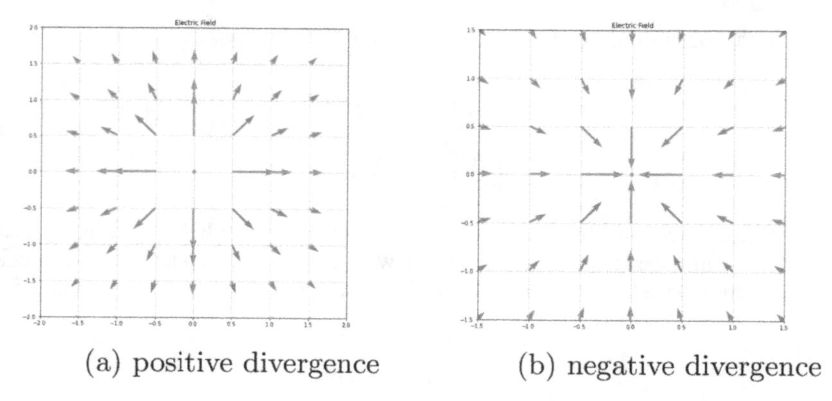

(a) positive divergence (b) negative divergence

Figure 3.6. Electric field divergence for a point charge, notice that the magnitude of the electric field decreases as we move away from the source.

An important difference between the integral and differential forms of Gauss's law is that while the integral form deals with the electric field over a surface, the differential form deals with the electric field at specific points.

Example 3.3. Divergence for the electric field created by a point charge.

Let's consider the electric field created by a point source. From equation (3.2), we can see that the magnitude of the electric field decreases by $1/r^2$, and is radial from the source.

We could apply the ∇ operator as defined by equation (3.10), by replacing $r^2 = x^2 + y^2 + z^2$, and apply the partial derivatives in Cartesian coordinates in each axis. An easier way, however, is to express the divergence in spherical coordinates as:

$$\nabla \cdot \mathbf{E} = \frac{1}{r}\frac{\partial}{\partial r}(r^2 E_r) + \frac{1}{\sin\theta}\frac{\partial}{\partial\theta}(E_\theta \sin\theta) + \frac{1}{r\sin\theta}\frac{\partial E_\phi}{\partial\phi} \tag{3.11}$$

Because the electric field only has radial components, the second and third terms of equation (3.11) will be zero.

$$\begin{aligned} \nabla \cdot \mathbf{E} &= \frac{1}{r}\frac{\partial}{\partial r}\left(r^2 \frac{q_{\text{enc}}}{4\pi\varepsilon_0 r^2}\right) \\ &= \frac{1}{r}\frac{\partial}{\partial r}\frac{q_{\text{enc}}}{4\pi\varepsilon_0} \\ &= 0 \end{aligned} \tag{3.12}$$

This result indicates that the divergence, caused by a point charge at the origin, at any point in space is zero. Wouldn't that be wrong? We know that there is a charge source at the origin, so there should be a divergence at the origin.

What is important to understand is that the electric field is undefined at the origin ($E \rightarrow \infty$ as $r \rightarrow 0$). So the function is not differentiable at the origin. In order to calculate the divergence at the origin, we will need to use a Dirac-delta function defined as:

$$\delta(x) = \begin{cases} \infty, & x = 0 \\ 0, & x \neq 0 \end{cases} \tag{3.13}$$

we also know that:

$$\int_{-\infty}^{+\infty} \delta(x)dx = 1 \tag{3.14}$$

and

$$\begin{aligned} \int_{-\infty}^{+\infty} f(x)\delta(x)dx &= f(0)\int_{-\infty}^{+\infty} \delta(x)dx \\ \int_{-\infty}^{+\infty} f(x)\delta(x)dx &= f(0) \end{aligned} \tag{3.15}$$

We will rewrite the electric field by changing the enclosed charge as the integral of the charge density over a volume V

$$\mathbf{E} = \frac{1}{4\pi\varepsilon_0}\int_V \frac{\hat{r}}{r^2}\rho(r')dV \tag{3.16}$$

taking the divergence on both sides of the equations:

$$\nabla \cdot \mathbf{E} = \frac{1}{4\pi\varepsilon_0}\int_V \nabla \cdot \frac{\hat{r}}{r^2}\rho(r')dV \tag{3.17}$$

from [2] we know that:

$$\nabla \cdot \frac{\hat{r}}{r^2} = 4\pi\delta^3(r - r') \tag{3.18}$$

So equation (3.17), can be expressed as:

$$\nabla \cdot \mathbf{E} = \frac{1}{4\pi\varepsilon_0}\int_V 4\pi\delta(r - r')\rho(r')dV \tag{3.19}$$

and finally from equations (3.15) and (3.19)

$$\frac{1}{4\pi\varepsilon_0}\int_V 4\pi\delta(r - r')\rho(r')dV = \frac{\rho(r)}{\varepsilon_0} \tag{3.20}$$

so,

$$\nabla \cdot \mathbf{E} = \frac{\rho(r)}{\varepsilon_0} \tag{3.21}$$

at the origin, in agreement with Gauss's law

Example 3.4. Given the electric vector field: $\mathbf{E} = \cos(3\pi x)\hat{i} + \sin\left(\frac{\pi}{3}y\right)\hat{j}$ (a) calculate the divergence at point $P(-4/3, 2)$, and (b) calculate the charge density.

First, we will plot the function of the given electric field to try to visualize the electric fields. We will use the function quiver in Python to generate this plot. The quiver function creates a 2D plot with arrows.

```
1 # Import required modules
2 import numpy as np
3 import matplotlib.pyplot as plt
4
5 # 1D arrays
6 x = np.arange(-5,5,0.5)
7 y = np.arange(-5,5,0.5)
8
9 # Meshgrid
10 X,Y = np.meshgrid(x,y)
11
12 # Assign vector directions
13
14 Ex = np.sin(3 * np.pi* X)
15 Ey = np.cos(np.pi/2 * Y)
16
17
18
19 # Plot Figure using Quiver function
20 plt.figure(figsize=(10, 10))
21
22 plt.quiver(X, Y, Ex, Ey, color='#A23BEC')
23 plt.plot(-1,3,'-or')
24 plt.title('Electric Field')
25
26 plt.grid()
27 plt.show()
```

Code 3.1. Python code to visualize Electric Field using function Quiver.

This gives us the following plot, where the red dot marks point P where we want to evaluate the divergence. We can see that at point P the electric field lines are moving towards this position, so it would be fair to assume that we will get a negative divergence. We can prove this by using equation (3.10) and the given electric field:

$$\begin{aligned} \nabla \cdot \mathbf{E} &= \left(\hat{i}\frac{\partial}{\partial x} + \hat{j}\frac{\partial}{\partial y} + \hat{k}\frac{\partial}{\partial z}\right) \cdot \left(\cos(3\pi x)\hat{i} + \sin\left(\frac{\pi}{3}y\right)\hat{j}\right) \\ &= -3\pi \sin(3\pi x) + \frac{\pi}{3}\cos\left(\frac{\pi}{3}y\right) \end{aligned} \tag{3.22}$$

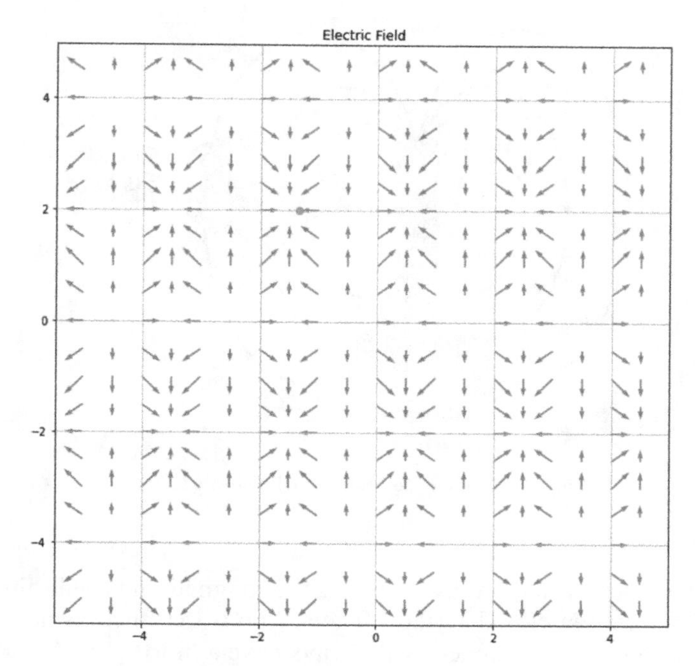

Figure 3.7. Plot of electric field.

Evaluating our solution at point $P(-4/3, 2)$ we obtain a divergence of $-\frac{\pi}{6}$ (figure 3.7).

To calculate the charge density we can use equation (3.9) substituting the left side of the equation with our previous results.

$$\rho = -\varepsilon_0 \frac{\pi}{6} \tag{3.23}$$

3.3 Gauss's law for magnetic fields

Once we are familiar with the first Maxwell's equation, understanding the second one is relatively easy. Gauss's law for the magnetic field also deals with a field going through a closed surface. However, we will see that no matter what kind of closed surface we define, the flux of the magnetic field will always be zero.

Gauss's law for magnetic fields in its integral form is defined as:

$$\oint_S (\mathbf{B} \cdot \hat{n}) ds = 0 \tag{3.24}$$

As before, the left side of the equation represents the net flux of a magnetic field across a closed surface, if it's equal to zero, it implies that all the magnetic field lines going into the closed surface will be equal to the magnetic field lines coming out of the closed surface.

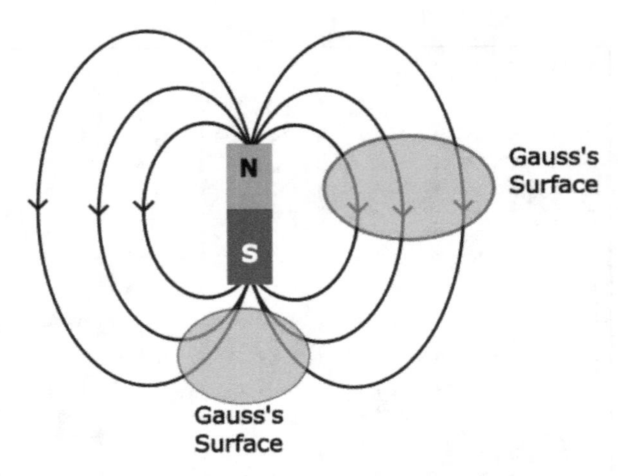

Figure 3.8. Magnetic field through Gauss's surface.

The question then is why? Why is it that the magnetic flux will always be zero regardless of our chosen closed surface? The reason is that there is no such thing as 'magnetic monopoles'. All magnets will always have a 'north' and 'south' pole, and the magnetic field will form loops from the north pole to the south pole.

In figure 3.8, we can see that regardless of the location of the Gaussian surface, the magnetic flux will always be zero.

Gauss's law for magnetic fields—differential form

The differential form of Gauss's law shows us that the divergence of the magnetic field at any point is zero.

$$\nabla \cdot \mathbf{B} = 0 \tag{3.25}$$

That can be easily understood when we remember that magnetic fields always create loops. So, regardless of the point in space that we chose, the number of magnetic field lines going into that point will be the same as the number of magnetic field lines coming out of the same point. This is not the case with electric fields, because an electric field can finish at an isolated negative charge, or flow away from a positive charge.

We can prove this by calculating the divergence of an arbitrary magnetic field.

Example 3.5. Calculate the divergence of a magnetic field caused by a current in a straight wire.

The magnetic field created by a current I traveling through an infinitely long wire is given by (3.26)

$$\mathbf{B} = \frac{\mu_o I}{2\pi\rho}\hat{\phi} \tag{3.26}$$

We need to express equation (3.25) in cylindrical coordinates to apply it to equation (3.26):

$$\nabla \cdot \mathbf{A} = \frac{1}{\rho}\frac{\partial}{\partial \rho}(\rho A_\rho) + \frac{1}{\rho}\frac{\partial A_\phi}{\partial \phi} + \frac{\partial A_z}{\partial z} \tag{3.27}$$

Applying (3.27) to (3.26) we obtain:

$$\nabla \cdot \mathbf{B} = \frac{1}{\rho}\frac{\partial}{\partial \phi}\frac{\mu_o I}{2\pi\rho} = 0 \tag{3.28}$$

This can be understood by looking at figure 3.9. We can see that the magnetic field **B** doesn't have any component on B_z, it will be reduced depending on how far you are from the wire, but the field itself will only have a B_ϕ component. Given a specific distance ρ away from the wire, the value of the magnetic field will be constant. So anywhere we measure, we will have a divergence equal to zero.

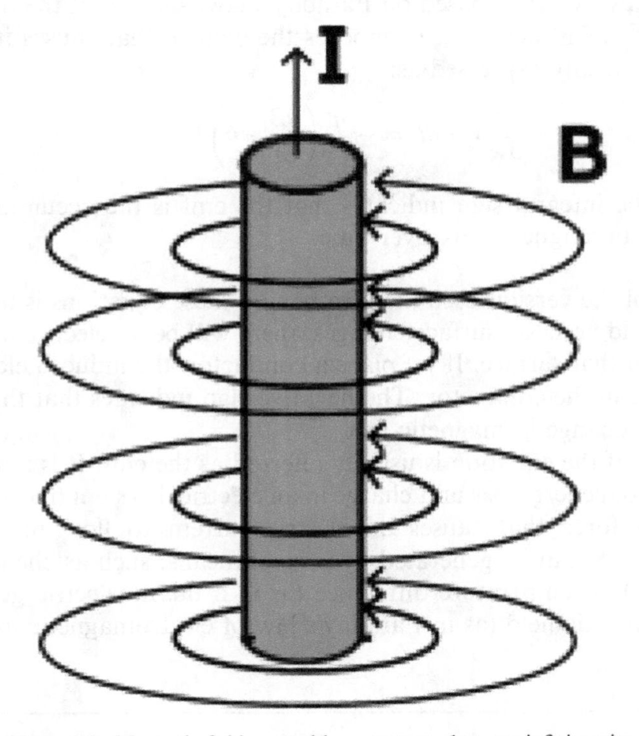

Figure 3.9. Magnetic field created by a current along an infinite wire.

3.4 Faraday's law

We will go ahead and describe the third Maxwell equation, Faraday's law.

Faraday's law of electromagnetic induction is a basic principle in electromagnetism that states that a change in the magnetic field within a closed loop of a conductor will induce an electromotive force (emf), or voltage, in the conductor. In other words, the law describes how a magnetic field can generate an electric current in a conductor. This principle forms the basis of generating electrical power in power plants and operating electric motors.

There are two equivalent equations that represent Faraday's law:

1. Faraday's law of electromagnetic induction, which is known as Faraday's first law, states that the emf induced in a conductor is equal to the rate of change of magnetic flux linked with the conductor:

$$\oint_S \mathbf{E} \cdot d\vec{l} = -\frac{d}{dt} \int_S (\mathbf{B} \cdot \hat{n}) ds \tag{3.29}$$

2. Lenz's law, which is based on Faraday's law, states that the direction of the induced emf is such that it opposes the change that causes it. This can be mathematically expressed as:

$$\oint_S \mathbf{E} \cdot d\vec{l} = -\int_S \left(\frac{\partial \mathbf{B}}{\partial t} \cdot \hat{n} \right) ds \tag{3.30}$$

   where the integral sign indicates that the emf is the accumulated effect of changes in magnetic flux over time.

Regardless of the version used the idea behind these equations is the same: when the magnetic field across a surface changes, there will be an electric field induced at the boundary of that surface. If we place a conductor, the induced electric field will create a current in the conductor. The negative sign indicates that the induced emf will oppose the change in magnetic flux.

The left side of the equation is usually referred as the emf. It is a measure of the electrical potential energy per unit charge in an electrical system (measured in volts). It is a driving force that causes an electric current to flow in a circuit. The electromotive force can be generated by several means, such as chemical reactions (as in a battery), a temperature difference (as in a thermoelectric generator), or a change in a magnetic field (as in Faraday's law of electromagnetic induction).

Example 3.6. A conducting bar moves to the right at a constant speed v, over some metallic rails into a constant magnetic field B, as shown in figure 3.10. Calculate the induced electric field and the direction of the produced current.

This is an interesting problem, because even though the magnetic field is constant, the area of the loop will change by moving the bar over the rails, and thus, the magnetic flux through this area.

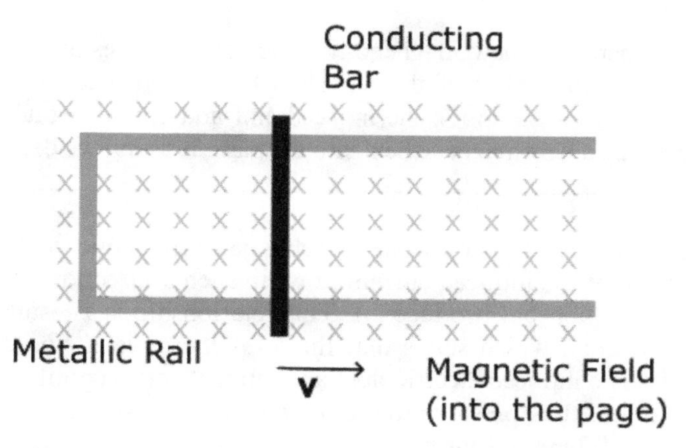

Figure 3.10. Conducting bar moving in a constant magnetic field.

We will use the right side of equation (3.29) to calculate the emf

$$\text{emf} = -\frac{d}{dt}\int_S (\mathbf{B} \cdot \hat{n})\, ds \tag{3.31}$$

Because the magnetic field is constant and perpendicular to the area form by the rails and the moving bar, we can simplify the previous equation to:

$$\begin{aligned}\text{emf} &= -\frac{d}{dt}B\int_S ds \\ &= -\frac{d}{dt}BA = \frac{d}{dt}BLx\end{aligned} \tag{3.32}$$

where A is the area of the loop, which can be expressed at the length of the rail times its position. From here we can see that the derivative of the position with respect to time is just the velocity of the rail, so we can calculate the emf as:

$$emf = -BLv \tag{3.33}$$

We can now calculate the induced electric field, by using the left side of equation (3.29) and equating it to our previous result.

$$\begin{aligned}\oint_S \mathbf{E} \cdot d\vec{l} &= -BLv \\ \mathbf{E}\oint_S d\vec{l} &= -BLv \\ E2(x + L) &= -Blv \\ E = &= -\frac{Blv}{2(x + L)}\end{aligned} \tag{3.34}$$

where the term $2(x + L)$ is the perimeter of the loop where the induced electric field is formed.

To determine the direction of the induced current, let's consider what happens when the rail moves to the right: the area of the loop increases, resulting in a greater magnetic flux, which induces an emf that causes a counter-clockwise current. This current then generates a new magnetic field out of the page, opposing the flux increase. On the other hand, if the wire had been moving left (and therefore decreasing the area of the loop), the induced emf and current would have been clockwise, creating a new magnetic field into the page, resisting the decrease in flux.

The induced electric field is opposite the change of magnetic field due to Lenz's law, which states that the induced current flows in such a direction as to oppose the change in magnetic flux that produced it. This relationship is a result of the law of conservation of energy, which states that the total energy in a closed system must remain constant. The induced electric field and current act to counteract the change in magnetic field and thus preserve the total energy in the system.

Faraday's law—differential form

The differential form of Faraday's Law can be written as:

$$\nabla \times \mathbf{E} = -\frac{\partial \mathbf{B}}{\partial t} \tag{3.35}$$

The cross-product between the ∇ operator and the electric field, commonly defined as the curl of the electric field, indicates that the electric field lines will tend to circulate in a close loop around a specific point. Therefore, the differential form of Faraday's law tells us that a time-varying magnetic field will create a circulating induced electric field.

It is important to mention that an induced electric field is different from the electric field produced by an electric charge. An electric field produced by an electric charge is a static field, meaning that it does not change over time unless the distribution of charges changes.

An induced electric field, on the other hand, is a dynamic field that arises due to a changing magnetic field. It is not caused by the presence of an electric charge, but rather by the changing magnetic field. The strength of the induced electric field depends on the rate of change of the magnetic field, as well as the characteristics of the conductor in which the field is induced, unlike an electric field produced by a charge.

It is worth remembering from previous sections, that the divergence of an electric field is the measure of how many sources (beginnings or ends) of electric field lines there are. Since an induced electric field forms closed loops, the divergence of this field is zero.

Example 3.7. Calculate the induced electric field created by the following time-varying magnetic field $\mathbf{B} = 4\cos(20t + 5x)\hat{k}$ Wb m^{-2}.

We will use Faraday's law as written in equation (3.35):

$$\begin{aligned} \nabla \times \mathbf{E} &= -\frac{\partial \mathbf{B}}{\partial t} \\ &= -\frac{\partial 4\cos(20t + 5x)}{\partial t}\hat{k} \\ &= -80\sin(20t + 5x)\hat{k} \end{aligned} \tag{3.36}$$

To solve for the left side of equation we will apply the ∇ in Cartesian coordinates:

$$\nabla \times \mathbf{E} = \begin{vmatrix} \hat{i} & \hat{j} & \hat{k} \\ \frac{\partial}{\partial x} & \frac{\partial}{\partial y} & \frac{\partial}{\partial z} \\ E_x & E_y & E_z \end{vmatrix} = \left(\frac{\partial E_z}{\partial y} - \frac{\partial E_y}{\partial z}\right)\hat{i} - \left(\frac{\partial E_z}{\partial x} - \frac{\partial E_x}{\partial z}\right)\hat{j} + \left(\frac{\partial E_y}{\partial x} - \frac{\partial E_x}{\partial y}\right)\hat{k} \tag{3.37}$$

Because equation (3.36) only has $\hat{k}$ components, we can simplify equation (3.37) to:

$$\left(\frac{\partial E_y}{\partial x} - \frac{\partial E_x}{\partial y}\right) = -80\sin(20t + 5x) \tag{3.38}$$

noticing that the right hand of the equation doesn't have any term depending on y or a constant value, we can set $E_x = 0$ to get:

$$\frac{\partial E_y}{\partial x} = -80\sin(20t + 5x) \tag{3.39}$$

To solve for E_y we integrate with respect of x:

$$E_y = -\int 80\sin(20t + 5x)dz = 40\cos(20t + 5x) \tag{3.40}$$

3.5 Ampère–Maxwell law

The Ampère–Maxwell Law is a mathematical equation that relates electric currents and magnetic flux. It was discovered by André-Marie Ampère and later modified by James Clerk Maxwell. The integral form of the equation states that electric and displacement currents are associated with a proportional magnetic field along any enclosing curve. The differential form of the equation states that the volume current density at any point in space is proportional to the spatial rate of change of the magnetic field and is perpendicular to the magnetic field at that point.

The equation is important because it explains how electricity and magnetism are related to each other. It also predicts that a rotating magnetic field occurs with a changing electric field. This is pivotal in showing that electromagnetic energy propagates as waves. The implications of the Ampère–Maxwell law are far-reaching. It is used to generate boundary conditions for waves that propagate through different materials. It is also used to explain the propagation of EM waves in matter and is an essential part of understanding the relationship between electricity and magnetism.

The integral form of the Ampère–Maxwell law is usually written as:

$$\oint_S \mathbf{B} \cdot d\vec{l} = \mu_0\left[I_{\text{enc}} + \varepsilon_0\frac{d}{dt}\int_S (\mathbf{E} \cdot \hat{n})ds\right] \tag{3.41}$$

where μ_0 is the free-space permeability ($4\pi \times 10^{-7}$ V s A m^{-1}), ε_0 is the free-space permittivity, and I_{enc} is the enclosed current. An enclosed current in the Ampère–Maxwell equation is a total current passing through a closed loop, and it is the sum of all currents that penetrate any surface bounded by the closed path.

The left side of equation (3.41) is very similar to the left side of the Faraday's law that we described in equation (3.29). As before, this side of the equation represents a field circulating around a close path. In the case of the Ampère–Maxwell law it is specifically a circulating magnetic field. The right hand of the equation indicates how can we create this circulating magnetic field. It can be created by an electric current and by a time-varying electric flux (we can find a similar equation on the right side of equation (3.29)).

One of Maxwell's contributions to Ampère's law was to account for the magnetic field created by time-varying electric fields. This simple term was key to understanding the link between magnetic and electric fields. Maxwell's term, known as the displacement current, states that a changing electric field creates an accompanying magnetic field. This is illustrated by Faraday's law of induction, which states that a changing magnetic field creates an accompanying electric field. Together, these two laws form the foundation of the theory of electromagnetism. This was a major breakthrough in understanding the behavior of electromagnetic fields and enabled the development of numerous technologies such as wireless communication, radar, and power transmission.

Example 3.8. Calculate the magnetic field created by a charging capacitor.

Let's imagine an RC circuit in which a voltage difference charges a capacitor over a period of time. Once the capacitor is charged current stops flowing in the circuit. What we would like to find is the magnetic field between the plates while the capacitor is charging.

In this case our Gaussian surface would look more like a butterfly net, with the end perpendicular to the magnetic field, as shown in figure 3.11. To calculate the rate of the electric flux, we need to understand that the electric field will be different inside the parallel plates than outside. From equation (3.1) we can define both regions as:

$$\frac{d}{dt}\left(\int_s \mathbf{E} \cdot \hat{n} ds\right) = \begin{cases} \dfrac{dQ}{dt}\dfrac{Qr^2}{\varepsilon_0 r_0^2}, & r \leqslant r_0 \\ \dfrac{d}{dt}\dfrac{Q}{\varepsilon_0}, & r > r_0 \end{cases} \tag{3.42}$$

where r is the radius of our Amperian loop, and r_0 is the radius of our capacitor plates.

From [1], we know that the capacitor's step response for an RC circuit will be:

$$v(t) = V(1 - \exp^{-t/\tau}) \tag{3.43}$$

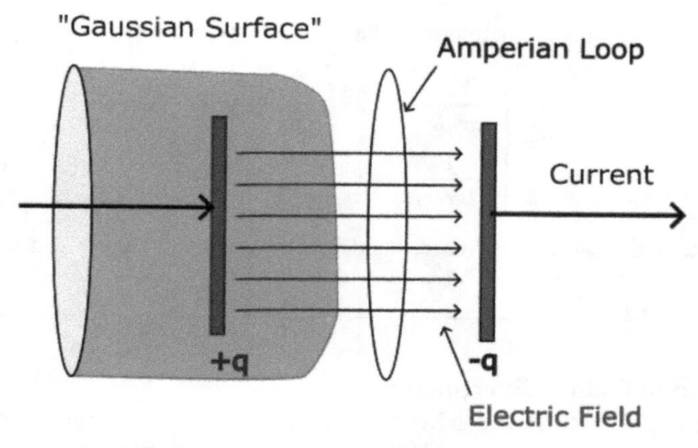

Figure 3.11. Electric field through parallel plate capacitor.

where τ is the response time of the circuit defined as $\tau = \mathrm{RC}$. Using $Q = \mathrm{CV}$ we can then find an expression of our capacitor's charge as a function of time:

$$q(t) = \mathrm{CV}(1 - \exp^{-t/\tau}) \tag{3.44}$$

Substituting equation (3.44) into (3.42), we obtained for the region $r > r_0$:

$$\begin{aligned} \frac{1}{\varepsilon_0}\frac{dQ}{dt} &= \frac{\mathrm{CV}}{\varepsilon_0}\frac{d}{dt}\left(1 - \exp^{-t/\mathrm{RC}}\right) \\ &= \frac{\mathrm{CV}}{\varepsilon_0}\left(\frac{1}{\mathrm{RC}}\exp^{-t/\mathrm{RC}}\right) \\ &= \frac{V}{\varepsilon_0 R}\exp^{-t/\mathrm{RC}} \end{aligned} \tag{3.45}$$

We can then define the rate of change of the electric flux Φ_E as:

$$\frac{d\Phi_E}{dt} = \begin{cases} \dfrac{V}{\varepsilon_0 R}\exp^{-t/\mathrm{RC}}\dfrac{r^2}{r_0^2}, & r \leqslant r_0 \\ \dfrac{V}{\varepsilon_0 R}\exp^{-t/\mathrm{RC}}, & r > r_0 \end{cases} \tag{3.46}$$

We can now apply the Ampère–Maxwell law as written in equation (3.41). We will assume $I_{\text{enc}} = 0$, due to the fact that no current travels between the plates.

$$\oint_S \mathbf{B} \cdot d\vec{l} = \mu_0\varepsilon_0\frac{d\Phi_E}{dt} \tag{3.47}$$

To solve for the left side of the previous equation, we will assume a loop of radius r,

$$\oint_S \mathbf{B} \cdot d\vec{l} = B(2\pi r) \tag{3.48}$$

And we get a solution for our magnetic field in both regions as:

$$B = \begin{cases} \dfrac{V}{2\pi\varepsilon_0 R}\exp^{-t/\mathrm{RC}}\dfrac{r}{r_0^2}, & r \leqslant r_0 \\ \dfrac{V}{2\pi\varepsilon_0 r R}\exp^{-t/\mathrm{RC}}, & r > r_0 \end{cases} \tag{3.49}$$

which is a magnetic field that increases inside the plates and decreases as we move away from our capacitor.

Ampère–Maxwell law—differential form

The differential form of the Maxwell–Ampère law is an equation that relates electric currents and magnetic flux. It states that the curl of the magnetic field at any point is equal to the current density there. This equation is derived from Ampère's law, which states that the circulation of the magnetic field around any imaginary closed loop is proportional to the current enclosed by the loop. The differential form of this law makes it possible to find the current distribution from a given magnetic field.

The Maxwell–Ampère law has several physical implications. It illustrates the speed of electromagnetic waves is the same as the speed of light, and it is used in understanding the principle of antennas. Additionally, it shows that the flow of electric current produces a magnetic field. The differential form of the Maxwell–Ampère law is expressed mathematically as:

$$\nabla \times \mathbf{B} = \mu_0\left(\vec{J} + \varepsilon_0\frac{\partial \mathbf{E}}{\partial t}\right) \tag{3.50}$$

The left side of the differential form of the Maxwell–Ampère law is represented by the curl of the magnetic field at any point. This equation states that the curl of the magnetic field at any point is equal to the current density there. In other words, the current density is a source for the curl of the magnetic field.

The right hand of the equation contains two terms; the first one is the electric current density. Electric current density is a vector quantity that measures the amount of electric current per unit area of cross-section. It is represented by the letter J and is measured in amperes per square meter (A m^{-2}). The direction of the current density is the same as the direction of the motion of the positive charges.

The second term is the displacement current density. Which is defined in terms of the rate of change of electric displacement field. It has the same units as electric current density, and it is a source of the magnetic field just as an actual current is. However, it is not an electric current of moving charges, but a time-varying electric field. In physical materials (as opposed to vacuum), there is also a contribution from the slight motion of charges bound in atoms, called dielectric polarization. Which will be explain further ahead in the next section.

Example 3.9. Given a magnetic field for a long coaxial cable carrying current, calculate the current density.

From [2, 3] we know that the magnetic field of a long coaxial cable as the one shown in figure 3.12 is given by:

$$\mathbf{B} = \begin{cases} \dfrac{\mu_0 I \rho}{2\pi a^2}\hat{\phi}, & \rho \leqslant a \\ \dfrac{\mu_0 I}{2\pi\rho}\hat{\phi}, & a < \rho \leqslant b \\ 0, & \rho > b \end{cases} \tag{3.51}$$

where I is the current flowing on the conductor, and ρ is the distance from the center of the cable, a is the radius of the inner wire, b the radius of the outer conductor. We want to calculate the current density using the differential form of the Ampère–Maxwell equation. We will calculate the curl of the magnetic field in cylindrical coordinates:

$$\nabla \times \mathbf{B} = \frac{1}{\rho}\begin{vmatrix} \hat{\rho} & \hat{\phi} & \hat{z} \\ \dfrac{\partial}{\partial\rho} & \dfrac{\partial}{\partial\phi} & \dfrac{\partial}{\partial z} \\ B_\rho & \rho B_\phi & B_z \end{vmatrix} \tag{3.52}$$

For our given **B** the curl is simplified to:

$$\nabla \times \mathbf{B} = \begin{cases} \dfrac{1}{\rho}\dfrac{\partial}{\partial\rho}\rho\dfrac{\mu_0 I \rho}{2\pi a^2}\hat{z} & \rho \leqslant a \\ \dfrac{1}{\rho}\dfrac{\partial}{\partial\rho}\rho\dfrac{\mu_0 I}{2\pi\rho}\hat{z}, & a < \rho \leqslant b \end{cases} \tag{3.53}$$

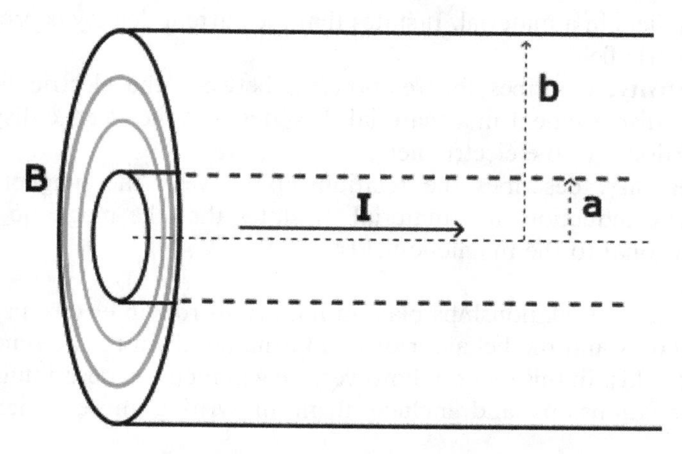

Figure 3.12. Coaxial cable with current I.

solving the partial derivatives, we get:

$$\nabla \times \mathbf{B} = \begin{cases} \frac{\mu_0 I}{\pi a^2}\hat{z}, & \rho \leqslant a \\ 0, & a < \rho \leqslant b \end{cases} \tag{3.54}$$

and we finally get an expression for our current density of:

$$J = \frac{I}{\pi a^2}\hat{z} \tag{3.55}$$

which is the current per unit area traveling through the inner conductor.

3.6 Constitutive relations

So far, we have assumed that our magnetic and electric fields were produced in a vacuum. We will now make some corrections to our Maxwell equations to include the effects of electromagnetic fields inside materials. For this, we will use what are called constitutive relations.

Maxwell's constitutive relationships are mathematical relationships that describe how an electric and magnetic field affects a material, and how a material affects an electric and magnetic field. In other words, they describe the way a material responds to electric and magnetic fields.

There are four Maxwell's constitutive relationships:

1. **Ohm's law:** describes the relationship between the current density and the electric field in a material. It states that the current density is proportional to the electric field.
2. **Conductivity:** describes the relationship between the current density and the electric field in a material. It states that the current density is proportional to the electric field.
3. **Permittivity:** describes the relationship between the electric field and the electric displacement in a material. It states that the electric displacement is proportional to the electric field.
4. **Permeability:** describes the relationship between the magnetic field and magnetic induction in a material. It states that the magnetic induction is proportional to the magnetic field.

These constitutive relationships play an important role in electromagnetism and are used to understand the behavior of various materials in the presence of electric and magnetic fields. In this section, however, we will focus on describing the last two constitutive relationships and include them in writing more general Maxwell equations.

We will first discuss **the electric displacement**, also known as the electric flux density, which is a measure of the flow of electric charge in a material. It is a vector quantity that is proportional to the total electric charge per unit area in a material. The electric displacement is denoted by the symbol **D** and is related to the electric field, E, by the relationship:

$$\mathbf{D} = \varepsilon\mathbf{E} = \varepsilon_0\mathbf{E} + \mathbf{P} \tag{3.56}$$

where ϵ is the electric permittivity of the material. The electric permittivity is a measure of how easily a material can become polarized in response to an electric field, and **P** is the electrical polarization.

Electrical polarization is the separation of electric charges within a material in response to an applied electric field. It is the process by which a material becomes charged when an electric field is applied to it. In an electrically neutral material, the positive and negative charges are evenly distributed. When an electric field is applied, the positive and negative charges begin to separate, leading to the creation of a dipole moment, as shown in figure 3.13. This dipole moment is proportional to the strength of the applied electric field and is a measure of the electrical polarization of the material.

The degree of polarization in a material depends on its electrical properties, such as its permittivity and conductivity. Some materials, such as insulators, are highly polarized in response to an electric field, while others, such as conductors, are not.

The electric displacement is important in electromagnetism as it is a key parameter in the calculation of electric fields and forces in dielectric materials. It also plays a crucial role in the understanding of electromagnetic waves and their interactions with materials.

The second constitutive relation to explore is the auxiliary field **H**. The auxiliary field H, also known as the magnetic field intensity or the magnetic field strength, is a measure of the magnetic field in a material. It is a vector quantity that is related to the magnetic induction, **B**, by the relationship:

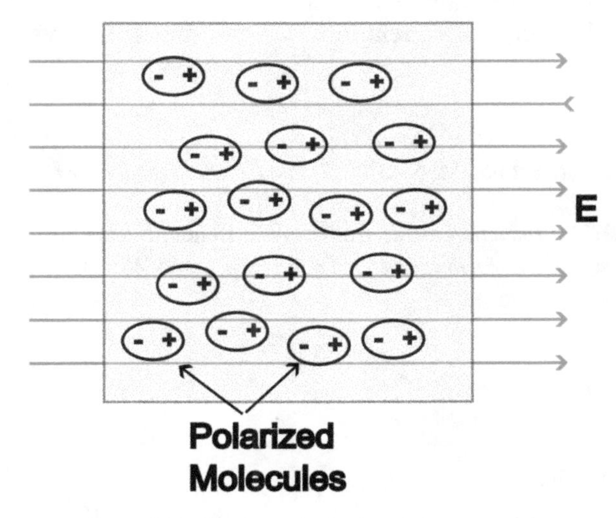

Figure 3.13. Electric polarization, **P** in a material.

Table 3.2. Maxwell's equations in differential form in a material.

Equation name	Differential form
Gauss's law of electric fields	$\nabla \cdot \mathbf{D} = \rho$
Gauss's law of magnetic fields	$\nabla \cdot \mathbf{B} = 0$
Faraday's law	$\nabla \times \mathbf{E} = -\frac{\partial \mathbf{B}}{\partial t}$
Ampère–Maxwell equation	$\nabla \times \mathbf{H} = \left(\mathbf{J} + \frac{\partial \mathbf{D}}{\partial t}\right)$

$$\mathbf{H} = \frac{1}{\mu}\mathbf{B} = \frac{1}{\mu_0}\mathbf{B} - \mathbf{M} \tag{3.57}$$

where μ is the magnetic permeability of the material. The magnetic permeability is a measure of how easily a material can become magnetized in response to a magnetic field, and **M** is the magnetization of the material.

Magnetization is a measure of the magnetic dipole moment per unit volume of a material. It is a vector quantity that represents the density of magnetic moments within a material and is proportional to the magnetic field within the material. The magnetization is denoted by the symbol **M**.

The magnetization of a material can be due to several sources, including the presence of permanent magnetic moments in the material, or the presence of induced magnetic moments in response to an external magnetic field. The magnetization of a material can be changed by applying a magnetic field, and the degree of magnetization is proportional to the strength of the applied magnetic field.

The auxiliary field **H** is important in electromagnetism as it is used to calculate the magnetic forces and torques in magnetic materials. It is also used in the calculation of magnetic fields in materials, and in the understanding of electromagnetic waves and their interactions with materials (table 3.2).

When adding the constitutive relations, we can find an expression for Maxwell equations in a material:

References

[1] Alexander C K and Sadiku M N O 2013 *Fundamentals of Electric Circuits* (New York: McGraw-Hill)

[2] Griffiths D J 2013 *Introduction to Electrodynamics* (London: Pearson)

[3] Pozar D 2005 *Microwave Engineering* 3rd edn (New York: Wiley)

IOP Publishing

Optical Sensors
An introduction with lab demonstrations
Victor Argueta-Diaz

Chapter 4

Electromagentic waves

4.1 Introduction

Before Maxwell's work, wave propagation was thought to be limited to mechanical systems, such as sound waves or vibrating strings, but Maxwell's equations showed that electric and magnetic fields, which are not mechanical systems, could also propagate in the form of waves. This revolutionized the way scientists thought about fields and opened up a whole new area of research. Maxwell's equations showed that electric and magnetic fields could interact with each other to produce electromagnetic waves, which could travel through empty space at a constant speed.

This prediction was confirmed by the detection of radio waves by German physicist Heinrich Hertz in the late 19th century. Hertz conducted a series of experiments in which he generated and detected radio waves, and showed that they followed the same laws of propagation and reflection as light.

These experiments provided strong experimental evidence for the validity of Maxwell's equations and established electromagnetism as a fundamental force in Nature, alongside gravity and the strong and weak nuclear forces. Since then, Maxwell's equations have been confirmed and verified by numerous experiments and are widely accepted as a cornerstone of modern physics.

In this chapter, we will use Maxwell's equations to explore the transmission of electromagnetic waves and their interaction with different media. We'll begin by deriving the wave equation from Maxwell's equations. We will then find expressions for Fresnel coefficients, that will describe the amount of reflection and transmission of light that occurs when it passes through the interface between two media with different refractive indices. We will look at specific cases of evanescent waves and the interaction of light with a metal boundary. We will conclude the chapter with a literature review of different sensors that have been implemented with the concepts described in this chapter.

doi:10.1088/978-0-7503-4876-8ch4

4.2 Electromagentic wave equation

We will first assume Maxwell's equations in free-space without any electric charge or current source:

$$\nabla \cdot \mathbf{E} = 0 \tag{4.1}$$

$$\nabla \cdot \mathbf{B} = 0 \tag{4.2}$$

$$\nabla \times \mathbf{E} = -\frac{\partial \mathbf{B}}{\partial t} \tag{4.3}$$

$$\nabla \times \mathbf{B} = \mu_0 \varepsilon_0 \frac{\partial \mathbf{E}}{\partial t} \tag{4.4}$$

We will now apply a curl to equation (4.3) as follows:

$$\nabla \times \nabla \times \mathbf{E} = \nabla \times \left(-\frac{\partial \mathbf{B}}{\partial t} \right) \tag{4.5}$$

Using the vector identity $\nabla \times \nabla \times \mathbf{E} = \nabla(\nabla \cdot \mathbf{E}) - \nabla^2 \mathbf{E}$ we obtain[1]:

$$\nabla(\nabla \cdot \mathbf{E}) - \nabla^2 \mathbf{E} = -\frac{1}{\partial t} \nabla \times \mathbf{B} \tag{4.6}$$

Substituting equations (4.1) and (4.4) into (4.6) to obtain:

$$\nabla^2 \mathbf{E} = \mu_0 \varepsilon_0 \frac{\partial^2 \mathbf{E}}{\partial t^2} \tag{4.7}$$

If we instead start with the curl of equation (4.4) we can obtain a similar equation for **B**:

$$\nabla^2 \mathbf{B} = \mu_0 \varepsilon_0 \frac{\partial^2 \mathbf{B}}{\partial t^2} \tag{4.8}$$

When comparing both expressions to the general form of a three-dimensional wave equation:

$$\nabla^2 A = \frac{1}{v^2} \frac{\partial^2 A}{\partial t^2} \tag{4.9}$$

We can then conclude that the **E** field (and the **B** field) propagates as a wave in free-space, with a constant speed of

$$v = \frac{1}{\sqrt{\mu_0 \varepsilon_0}} \approx 3 \times 10^8 \text{ m s}^{-1} \tag{4.10}$$

[1] In appendix A, the reader can find some useful calculus vector identities.

The wave equation, shown in equation (4.7), is a partial differential equation that describes the propagation of waves in a variety of physical systems, including mechanical waves, electromagnetic waves, and acoustic waves.

The solutions to the wave equation depend on the boundary conditions and initial conditions of the physical system. For a given system, there are many possible wave solutions that can be used to describe the behavior of the system. Some common solutions to the wave equation include:

1. **Sinusoidal waves**: these are wave solutions in the form of sinusoidal functions that describe simple harmonic motion. They are often used to describe the propagation of waves in strings and other linear systems.
2. **Gaussian waves**: these are wave solutions in the form of Gaussian functions that describe waves that spread out in space over time. They are often used to describe the propagation of light and other electromagnetic waves.
3. **Traveling waves**: these are wave solutions that move through space without changing shape or amplitude.
4. **Standing waves**: these are wave solutions that remain in a fixed position and have nodes and antinodes. They are often used to describe the behavior of waves in confined systems, such as pipes and cavities.

There are many other possible solutions to the wave equation, and the choice of solution depends on the particular physical system being studied and the desired level of accuracy and complexity of the solution.

We will work first with a traveling wave propagating in free-space. These waves can be described by a plane wave solution of the form[2]:

$$\mathbf{E}(\mathbf{r}, \mathbf{t}) = \mathbf{E}_0 \exp\left(-j(\mathbf{k} \cdot \mathbf{r} - \omega t)\right) \tag{4.11}$$

where $\mathbf{E}_0$ is a constant complex amplitude, ω is the angular frequency, $\mathbf{k}$ is a vector whose magnitude is ω/c (as we will prove later on), and its direction is the same as the direction of propagation of the wave, and $\mathbf{r}$ is a position vector. Notice that we are using a phasor notation for our electric field (table 4.1). One advantage of working with phasors is that we can rewrite Maxwell's equations in what is known as their time-harmonic form:where the s subindices indicate the phasor form of the field. If we use the time-harmonic equations and apply the same process that we described at the beginning of this section to obtain the wave equation (equations (4.5) to (4.8), we will obtain the following expressions:

$$\nabla^2\mathbf{E} + \omega^2\mu_0\varepsilon_0\mathbf{E} = 0 \tag{4.12}$$

and

$$\nabla^2\mathbf{H} + \omega^2\mu_0\varepsilon_0\mathbf{H} = 0 \tag{4.13}$$

also known as Helmholtz equations.

[2] Even though it is usual to refer to $i = \sqrt{-1}$, my electrical engineering background keeps pushing me to write $j = \sqrt{-1}$, I hope the reader and all physicists will forgive me.

Table 4.1. Maxwell's equations in time-harmonic form.

Time-harmonic form
$\nabla \cdot \mathbf{D}_s = \rho_s$
$\nabla \cdot \mathbf{B}_s = 0$
$\nabla \times \mathbf{E}_s = -j\omega\mathbf{B}_s$
$\nabla \times \mathbf{H}_s = \mathbf{J}_s + j\omega\mathbf{D}_s$

Let's substitute equation (4.11) into equation (4.12):

$$\nabla^2\mathbf{E}_0 e^{-j\mathbf{k}\cdot\mathbf{r}} + \omega^2\mu_0\varepsilon_0\mathbf{E}_0 e^{-j\mathbf{k}\cdot\mathbf{r}} = 0 \tag{4.14}$$

where we eliminate the time dependent factor on the electric field. Let's also assume that the wave is propagating in the z-direction, that is: $\mathbf{r} = z\hat{z}$

$$\mathbf{E}_0\nabla^2 e^{-j\mathbf{k}z} + \omega^2\mu_0\varepsilon_0\mathbf{E}_0 e^{-j\mathbf{k}z} = 0 \tag{4.15}$$

Calculating the second derivative of $\exp -j\mathbf{k}z$

$$-k^2\mathbf{E}_0 e^{-j\mathbf{k}z} + \omega^2\mu_0\varepsilon_0\mathbf{E}_0 e^{-j\mathbf{k}z} = 0 \tag{4.16}$$

which can be simplified to just:

$$-k^2 + \omega^2\mu_0\varepsilon_0 = 0 \tag{4.17}$$

which leads to the relation:

$$k = \omega\sqrt{\mu_0\varepsilon_0} = \frac{\omega}{c} \tag{4.18}$$

we can find an interesting relationship between **E** and our wave vector **k**, by applying Gauss's law in free-space in the absence of free charges (i.e. $\rho_s = 0$)

$$\begin{gathered}\nabla \cdot \mathbf{E} = 0 \\ \nabla \cdot \mathbf{E}_0 e^{-j\mathbf{k}z} = 0 \\ \mathbf{E}_0 \cdot (-j\mathbf{k})e^{-j\mathbf{k}z} = 0 \\ (\mathbf{E}_0 \cdot \mathbf{k})(e^{-j\mathbf{k}z}) = 0\end{gathered} \tag{4.19}$$

from the first term of equation (4.19), we conclude that:

$$\mathbf{E}_0 \cdot \mathbf{k} = 0 \tag{4.20}$$

which implies that **E** is perpendicular to the wavenumber **k**.

Finally, by using Faraday law's, we can find our magnetic field:

$$\begin{gathered}\nabla \times \mathbf{E} = -j\omega\mu_0\mathbf{H} \\ \mathbf{H} = \frac{1}{\eta}\mathbf{k} \times \mathbf{E}\end{gathered} \tag{4.21}$$

where $\eta = \sqrt{\omega_0/\varepsilon_0}$ and is known as the wave impedance, with an approximate value of 377 Ω for free-space (figure 4.1). In this chapter, we will explore the propagation of EM waves. we will use either the time-harmonic form or the time-varying Maxwell equations as necessary. For now, however, we will make a small comment regarding the speed of propagation. If we assume a lossless material, without any free charges, or free currents, then our speed of propagation in such medium will be given by:

$$v = \frac{1}{\sqrt{\varepsilon\mu}} \tag{4.22}$$

where ϵ is the permittivity of the medium and can be calculated from $\varepsilon = \varepsilon_r\varepsilon_0$. where ε_r is the relative permittivity of the medium. Similarly, μ is the permeability of the medium, and can be calculated from $\mu = \mu_r\mu_0$, where μ_r is the relative permeability of the medium, so:

$$v = \frac{1}{\sqrt{\varepsilon_r\varepsilon_0\mu_r\mu_0}} = \frac{c}{\sqrt{\varepsilon_r\mu_r}} \tag{4.23}$$

where c is the speed of light as expressed in equation (4.10). For most material, $\mu_r \approx 1$, so finally, we can define the index of refraction of a material as:

$$n = \sqrt{\varepsilon_r} \tag{4.24}$$

and the speed of propagation of light in a lossless material as

$$v = \frac{c}{n} \tag{4.25}$$

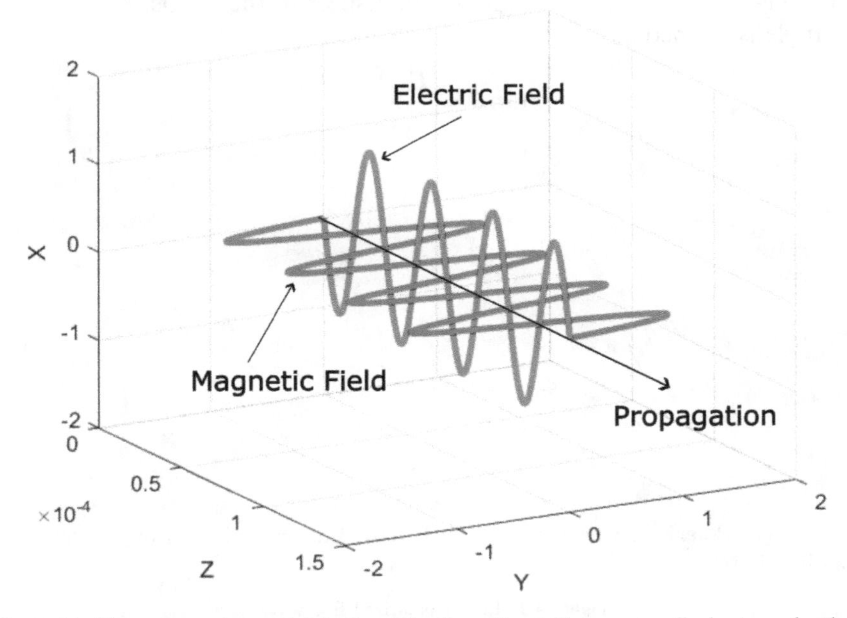

Figure 4.1. Plane wave propagating in the z-direction. E and H are perpendicular to each other.

4.2.1 Polarization

We will now look a little bit closer at the orientation of the electric field. This orientation is referred to as the polarization of the wave. Considering a plane wave traveling along the z-axis of the form:

$$\mathbf{E}(\mathbf{z}, \mathbf{t}) = \mathbf{E}_0 \cos(kz - \omega t) \tag{4.26}$$

From equation (4.20), we know that $\mathbf{E}_0$ can be pointing in any direction perpendicular to $\mathbf{k}$. So two of our possible waves are:

$$E_x(z, t) = E_{0x} \cos(kz - \omega t)\hat{x} \tag{4.27}$$

and,

$$E_y(z, t) = E_{0y} \cos(kz - \omega t + \delta)\hat{y} \tag{4.28}$$

where δ is the relative phase difference between our two waves, and E_{0x} and E_{0y} are real amplitudes. Notice that each wave oscillates only in either the x-direction or the y-direction. So, we say that each of these waves is linearly polarized. This is shown in figure 4.2.

If we assume that both waves are in phase, $\delta = 0$ or a multiple of $\pm 2\pi$, then we can express the sum of both waves as:

$$\mathbf{E}(z, t) = \left(E_{0x}\hat{x} + E_{0y}\hat{y}\right) \cos(kz - \omega t) \tag{4.29}$$

This $\mathbf{E}$ field will be oscillating with a constant amplitude, $\left(E_{0x}\hat{x} + E_{0y}\hat{y}\right)$, so this wave will also be linearly polarized. If we were able to observe the electric field traveling toward us, shown in figure 4.2, we would see the electric field oscillating on a straight line in the xy-plane tilted at an angle θ with respect to the x-axis.

This angle is defined as:

$$\tan \theta = \frac{E_{0y}}{E_{0x}} \tag{4.30}$$

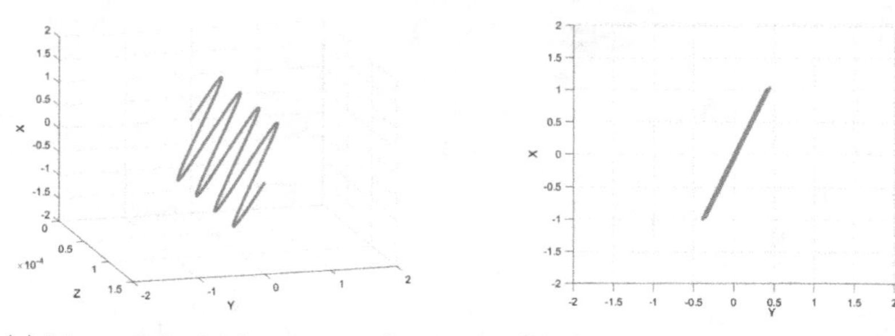

(a) Linear Polarized wave traveling in the z-direction

(b) Linear Polarization, Front view

Figure 4.2. Linear polarized EM wave.

Example 4.1. Assume two orthogonal plane waves, with equal amplitude, traveling in the z-direction, with a phase difference, $\delta = \pi/2$. Describe the net electric field.

We begin by writing an expression for each orthogonal electric field:

$$E_x(z,\, t) = E_0 \cos(kz - \omega t)\hat{x} \tag{4.31}$$

$$E_y(z,\, t) = E_0 \cos\left(kz - \omega t + \frac{\pi}{2}\right)\hat{y} \tag{4.32}$$

let's evaluate each field at a point in space ($z = 0$) and see how it changes with time.

$$E_x = E_0 \cos(\omega t) \tag{4.33}$$

$$E_y = E_0 \cos\left(\omega t + \frac{\pi}{2}\right) = -E_0 \sin(\omega t) \tag{4.34}$$

at $t = 0$, **E** lies along the x-axis. at a later time, $t = \pi/4\omega$ the **E** will lie 45° above the x-axis, at $t = \pi/2\omega$, it will lie on the y-axis. We can see how this electric field will travel counter-clockwise in a circular pattern. This is also known as left-circularly polarized. A representation of this wave traveling in space is shown in figure 4.3.

Linear and circular polarizations are special cases of elliptical polarization. In fact, any polarization state that can be represented as a combination of two perpendicular linear polarizations (e.g., horizontal and vertical) can be expressed as an elliptical polarization state. In an elliptically polarized wave, the electric field vector traces out an elliptical path as it propagates through space. The shape and orientation of the elliptical path depend on the relative amplitudes and phases of the two perpendicular linear polarizations that make up the wave.

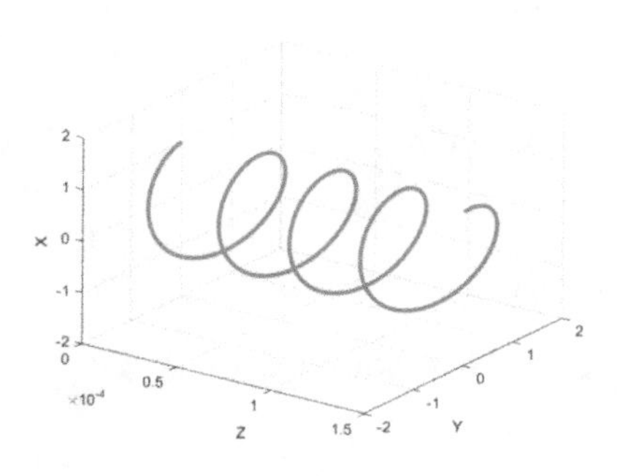

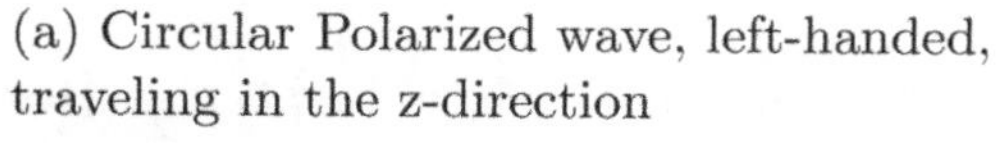

(a) Circular Polarized wave, left-handed, traveling in the z-direction

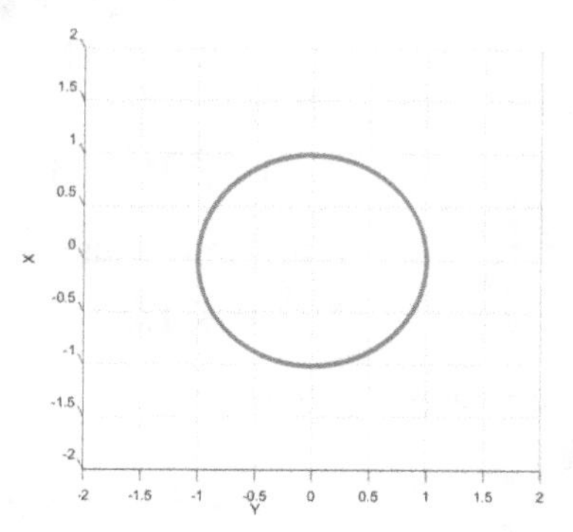

(b) Circular Polarization, Front view

Figure 4.3. Circular polarized EM wave.

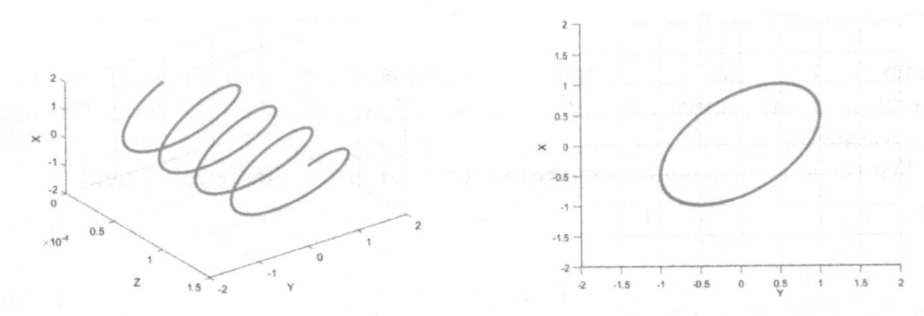

(a) Elliptical Polarized wave traveling in the z-direction

(b) Elliptical Polarization, Front view

Figure 4.4. Elliptical polarized EM wave.

When the amplitudes of the two linear polarizations are equal and out of phase by 90°, the resulting polarization is circular. When the two polarizations are in phase, the resulting polarization is linear.

Elliptical polarization is a more general polarization state that includes both linear and circular polarizations as special cases (figure 4.4).

4.2.2 Poynting vector

The Poynting vector, named after English physicist John Henry Poynting, is a mathematical concept used in electromagnetism to describe the flow of energy in electromagnetic fields.

To obtain an expression for the Poynting vector we will start with Ampère's law and Faraday's law in free-space without any charge or current source as written in equations (4.3) and (4.4). We first add the dot product of **E** to equation (4.4):

$$\mathbf{E} \cdot (\nabla \times \mathbf{H}) = \mathbf{E} \cdot \varepsilon_0 \frac{\partial \mathbf{E}}{\partial t} \tag{4.35}$$

Using the vector identity:

$$\nabla \cdot (\mathbf{A} \times \mathbf{B}) = \mathbf{B} \cdot (\nabla \times \mathbf{A}) - \mathbf{A} \cdot (\nabla \times \mathbf{B}) \tag{4.36}$$

we can rewrite our previous equation as:

$$\nabla \cdot (\mathbf{H} \times \mathbf{E}) = \mathbf{E} \cdot (\nabla \times \mathbf{H}) - \mathbf{H} \cdot (\nabla \times \mathbf{E}) \tag{4.37}$$

We are now going to calculate the dot product of **H** to equation (4.3):

$$\mathbf{H} \cdot (\nabla \times \mathbf{E}) = \mathbf{H} \cdot -\mu_0 \frac{\partial \mathbf{H}}{\partial t} \tag{4.38}$$

Substituting equations (4.35) and (4.38) into equation (4.37), and replacing $(\mathbf{H} \times \mathbf{E})$ by $-(\mathbf{E} \times \mathbf{H})$ we obtain:

$$\begin{aligned}\nabla \cdot (\mathbf{E} \times \mathbf{H}) &= -\mathbf{E} \cdot \varepsilon_0 \frac{\partial \mathbf{E}}{\partial t} - \mathbf{H} \cdot -\mu_0 \frac{\partial \mathbf{H}}{\partial t} \\ &= -\frac{1}{2}\varepsilon_0 \frac{\partial E^2}{\partial t} - \frac{1}{2}\mu_0 \frac{\partial H^2}{\partial t} \\ &= -\frac{1}{2}\frac{\partial}{\partial t}\left(\varepsilon_0 E^2 + \mu_0 H^2\right)\end{aligned} \tag{4.39}$$

The terms on the right side of the previous equation represent the energy per volume stored in the electric and magnetic fields, while the left side is the divergence of the Poynting vector usually defined as:

$$\mathbf{S} = \mathbf{E} \times \mathbf{H} \tag{4.40}$$

Which represents the power per unit area in W m^{-2} and represents the instantaneous power density associated with an EM field at a given point in space. Integrating the Poynting vector over a closed surface gives the net power flowing out of that surface.

The Poynting vector has both magnitude and direction. Its direction is perpendicular to both the electric and magnetic fields, and points in the direction of energy flow. The magnitude of the Poynting vector gives the amount of energy per unit area per unit time that is flowing through a surface perpendicular to the direction of the vector.

4.3 Fresnel coefficients: reflection at an interface

With the definitions described in the previous section for a plane wave, we can now calculate the Fresnel coefficients. The Fresnel coefficients are mathematical expressions that describe the behavior of light waves at the boundary between two different materials. The Fresnel coefficients are used to calculate the amount of light that is reflected and transmitted at the boundary. They depend on the materials at the interface, on the angle of incidence, and the polarization of light.

There are two Fresnel coefficients: the reflection coefficient (Γ) and the transmission coefficient (T). The reflection coefficient describes the proportion of the incident light that is reflected at the interface, while the transmission coefficient describes the proportion of the incident light that is transmitted through the interface. The Fresnel coefficients are complex numbers, meaning that they have both magnitude and phase information. The magnitude of the Fresnel coefficients determines the amount of light that is reflected or transmitted, while the phase of the Fresnel coefficients determines the phase shift of the light as it is reflected or transmitted.

Let's assume a plane wave that is incident at an angle θ_i at the boundary of two media that are characterized by their refractive index, as shown in figure 4.5. The plane containing both the normal to the surface and the direction of propagation of the incident wave is known as the plane of incidence. In our case we will be assuming that the wave propagates in the z-direction, and that our plane of incidence will be the yz plane (the x-axis will be pointing into the page).

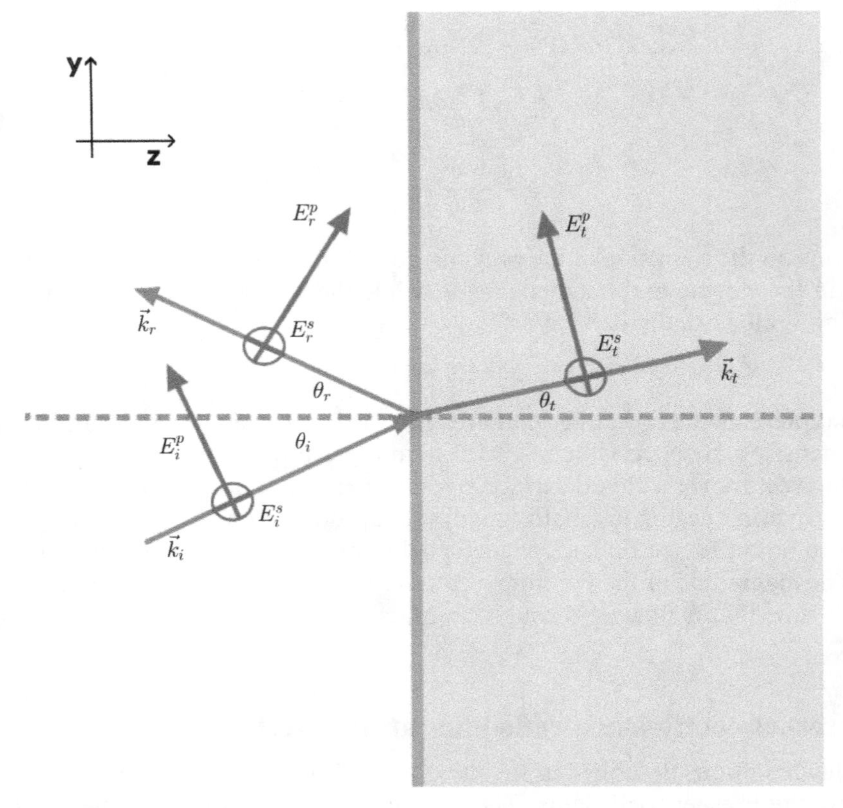

Figure 4.5. Plane wave incident at an angle $\theta_i s$, at the interface between two different materials. Image created by L Argueta-Slaughter.

4.3.1 *S*-polarization

We will consider the case of linearly polarized light with its electric field perpendicular to the plane of incidence, that is, pointing into the page for our example. This particular kind of polarization is known as an *s*-polarized wave (or TE polarization). The wave numbers **k**, can then be written as:

$$\begin{aligned}\mathbf{k}_{\mathrm{i}} &= k_{\mathrm{i}}(\sin\theta_{\mathrm{i}}\hat{y} + \cos\theta_{\mathrm{i}}\hat{z})\\ \mathbf{k}_{\mathrm{r}} &= k_{\mathrm{r}}(\sin\theta_{\mathrm{r}}\hat{y} - \cos\theta_{\mathrm{r}}\hat{z})\\ \mathbf{k}_{\mathrm{t}} &= k_{\mathrm{t}}(\sin\theta_{\mathrm{t}}\hat{y} + \cos\theta_{\mathrm{t}}\hat{z})\end{aligned} \tag{4.41}$$

and the electric field as:

$$\begin{aligned}\mathbf{E}_{\mathrm{i}} &= \hat{x}E_{\mathrm{i}}\exp j[k_{\mathrm{i}}(y\sin\theta_{\mathrm{i}} + z\cos\theta_{\mathrm{i}}) - \omega_{\mathrm{i}}t]\\ \mathbf{E}_{\mathrm{r}} &= \hat{x}E_{\mathrm{r}}\exp j[k_{\mathrm{r}}(y\sin\theta_{\mathrm{r}} - z\cos\theta_{\mathrm{r}}) - \omega_{\mathrm{r}}t]\\ \mathbf{E}_{\mathrm{t}} &= \hat{x}E_{\mathrm{t}}\exp j[k_{\mathrm{t}}(y\sin\theta_{\mathrm{t}} + z\cos\theta_{\mathrm{t}}) - \omega_{\mathrm{t}}t]\end{aligned} \tag{4.42}$$

We now need to apply boundary conditions for Maxwell equations. These conditions require that the E components that are parallel to the interface should be continuous on both side of the interface (i.e. at $z = 0$). Furthermore, at the interface we need for the electric field one the left side of the interface to be equal to the electric field on the right side. That is: $\mathbf{E}_i + \mathbf{E}_r = \mathbf{E}_t$. This can be expressed as:

$$\begin{aligned} &\hat{x}E_i \exp j[k_i(y \sin\theta_i) - \omega_i t] + \\ &\hat{x}E_r \exp j[k_i(y \sin\theta_r) - \omega_r t] \\ &= \hat{x}E_t \exp j[k_i(y \sin\theta_t) - \omega_t t] \end{aligned} \tag{4.43}$$

This equation must be valid for any values of t and y, which implies that the angular frequency for all waves should be the same, $\omega_i = \omega_r = \omega_t s$, and:

$$k_i \sin\theta_i = k_r \sin\theta_r = k_t \sin\theta_t \tag{4.44}$$

From equation (4.18) we can rewrite the wavenumber in terms of the refractive index of the material as $k = n\omega/c$, and we can obtain two very well known optical formulas: the law of reflection and Snell's law

$$\begin{aligned} n_1\frac{\omega_i}{c}\sin\theta_i &= n_1\frac{\omega_r}{c}\sin\theta_r \\ \theta_i &= \theta_r \end{aligned} \tag{4.45}$$

and

$$\begin{aligned} n_1\frac{\omega_i}{c}\sin\theta_i &= n_2\frac{\omega_t}{c}\sin\theta_t \\ n_1\sin\theta_i &= n_2\sin\theta_t \end{aligned} \tag{4.46}$$

Because the exponents are all equal to each other, equation (4.43) can be reduced to:

$$E_i + E_r = E_t \tag{4.47}$$

Now that we have defined some boundary conditions for our $\mathbf{E}$ field, we will need to find the boundary conditions for our $\mathbf{H}$ field. We will use equation (4.21) to obtain our magnetic field:

$$\begin{aligned} \mathbf{H}_i &= \frac{n_1}{c}[E_i(-\hat{z}\sin\theta_i + \hat{y}\cos\theta_i)]\exp j[k_i(y\sin\theta_i + z\cos\theta_i) - \omega_i t] \\ \mathbf{H}_r &= \frac{n_1}{c}[E_r(-\hat{z}\sin\theta_r - \hat{y}\cos\theta_r)]\exp j[k_r(y\sin\theta_i - z\cos\theta_i) - \omega_r t] \\ \mathbf{H}_t &= \frac{n_2}{c}[E_t(-\hat{z}\sin\theta_t + \hat{y}\cos\theta_t)]\exp j[k_t(y\sin\theta_t + z\cos\theta_t) - \omega_t t] \end{aligned} \tag{4.48}$$

We need the parallel components of the magnetic field to be the same at both sides of the interface. That is the y-components of the $\mathbf{H}$ need to be continuous at $z = 0$.

$$\frac{n_1}{c}[E_i(\hat{y}\cos\theta_i)] + \frac{n_1}{c}[E_r(-\hat{y}\cos\theta_r)] = \frac{n_2}{c}[E_t(\hat{y}\cos\theta_t)] \tag{4.49}$$

where we use the law of reflection and assuming that the exponents will all be equal to each other as before. Equation (4.49) simplifies to:

$$n_1(E_i - E_r)\cos\theta_i = n_2 E_t \cos\theta_t \tag{4.50}$$

To calculate the reflection coefficient we will use equations (4.47) and (4.50) as follows:

$$\begin{aligned} E_i + E_r &= E_t \\ E_i - E_r &= \frac{n_2 \cos\theta_t}{n_1 \cos\theta_i} E_t \end{aligned} \tag{4.51}$$

and add both equations together to obtain:

$$2E_i = \left[1 + \frac{n_2 \cos\theta_t}{n_1 \cos\theta_i}\right] E_t \tag{4.52}$$

solving for the ration between E_t/E_i, we obtain

$$T_s = \frac{E_t}{E_i} = \frac{2n_1 \cos\theta_i}{n_1 \cos\theta_1 + n_2 \cos\theta_t} \tag{4.53}$$

similarly, to obtain the ratio of the reflected beam with the incident beam, we subtract the expressions in equation (4.51) to obtain:

$$2E_r = \left[1 - \frac{n_2 \cos\theta_t}{n_1 \cos\theta_i}\right] E_t \tag{4.54}$$

Dividing equation (4.54) by equation (4.52) we obtain:

$$\Gamma_s = \frac{E_r}{E_i} = \frac{n_1 \cos\theta_i - n_2 \cos\theta_t}{n_1 \cos\theta_i + n_2 \cos\theta_t} \tag{4.55}$$

If we substitute Snell's equation into equations (4.53) and (4.55), we can obtain an expression for both coefficients only in terms of θ_1, and the refractive index of the materials.

$$\Gamma_s = \frac{n_1 \cos\theta_i - n_2\sqrt{1 - \left(\frac{n_1}{n_2}\sin\theta_i\right)^2}}{n_1 \cos\theta_i + n_2\sqrt{1 - \left(\frac{n_1}{n_2}\sin\theta_i\right)^2}}, \tag{4.56}$$

and

$$T_s = \frac{2n_1 \cos\theta_i}{n_1 \cos\theta_i + n_2\sqrt{1 - \left(\frac{n_1}{n_2}\sin\theta_i\right)^2}}, \tag{4.57}$$

Example 4.2. An electromagnetic wave having an amplitude of 1.0 V m^{-1} arrives at an angle of 50.0° to the normal in air on a glass plate of index 1.45. The wave's electric field is entirely perpendicular to the plane of incidence. Determine the amplitude of the reflected and transmitted wave.

Because we have our electric field perpendicular to the plane of incidence, we will need the Fresnel coefficients for an *s*-polarized wave, that is equations (4.56) and (4.57).

$$\Gamma_s = \frac{n_1 \cos\theta_i - n_2\sqrt{1 - \left(\frac{n_1}{n_2}\sin\theta_i\right)^2}}{n_1 \cos\theta_i + n_2\sqrt{1 - \left(\frac{n_1}{n_2}\sin\theta_i\right)^2}},$$

and

$$T_s = \frac{2n_1\cos\theta_i}{n_1 \cos\theta_i + n_2\sqrt{1 - \left(\frac{n_1}{n_2}\sin\theta_i\right)^2}},$$

for $\theta_1 = 50°$, $n_1 = 1$, and $n_2 = 1.45$, we obtain

$$\Gamma_s = \frac{\cos 50 - 1.45\sqrt{1 - \left(\frac{1}{1.45}\sin 50\right)^2}}{\cos 50 + 1.45\sqrt{1 - \left(\frac{1}{1.45}\sin 50\right)^2}}$$

and we obtain a value of $\Gamma_s = -0.3944$, so our reflected electric field will be

$$\begin{aligned} E_r &= \Gamma_s E_i \\ &= -0.3944 \text{ V m}^{-1} \end{aligned} \tag{4.58}$$

Similarly, the transmission coefficient will be given by:

$$\begin{aligned} T_s &= \frac{2\cos 50}{\cos 50 + 1.45\sqrt{1 - \left(\frac{1}{1.45}\sin 50\right)^2}} \\ &= 0.8444 \end{aligned} \tag{4.59}$$

4.3.2 *P*-polarization

Now that we have defined the Fresnel coefficients for an *s*-polarized wave, we will derive the Fresnel coefficients for *p*-polarized (or TM polarization) wave where the electric field is parallel to the plane of incidence.

Our wavenumber **k** will still be defined by the same expressions from equation (4.6). Our electric field, however, will now be defined by:

$$\begin{aligned}\mathbf{E}_\mathrm{i} &= E_\mathrm{i}(\hat{y}\cos\theta_\mathrm{i} - \hat{z}\sin\theta_\mathrm{i})\exp j[k_\mathrm{i}(y\sin\theta_\mathrm{i} + z\cos\theta_\mathrm{i}) - \omega_\mathrm{i}t]\\ \mathbf{E}_\mathrm{r} &= E_\mathrm{r}(\hat{y}\cos\theta_\mathrm{r} + \hat{z}\sin\theta_\mathrm{r})\exp j[k_\mathrm{r}(y\sin\theta_\mathrm{r} - z\cos\theta_\mathrm{r}) - \omega_\mathrm{r}t]\\ \mathbf{E}_\mathrm{t} &= E_\mathrm{t}(\hat{y}\cos\theta_\mathrm{t} - \hat{z}\sin\theta_\mathrm{t})\exp j[k_\mathrm{t}(y\sin\theta_\mathrm{t} + z\cos\theta_\mathrm{t}) - \omega_\mathrm{t}t]\end{aligned} \tag{4.60}$$

As before, we will need to apply boundary conditions so that parallel components to the interface of the **E** field are continuous at $z = 0$, for our p-polarized light, this will be the $\hat{y}$ component of the field.

$$\begin{aligned}E_\mathrm{i}\cos\theta_\mathrm{i}\hat{y} + E_\mathrm{r}\cos\theta_\mathrm{r}\hat{y} &= E_\mathrm{t}\cos\theta_\mathrm{t}\hat{y}\\ (E_\mathrm{i} + E_\mathrm{r})\cos\theta_\mathrm{i} &= E_\mathrm{t}\cos\theta_\mathrm{t}\end{aligned} \tag{4.61}$$

where we have assumed that the phase components will be equal as we derived before. We will now find an expression for **H** using equation (4.21) and apply boundary conditions to these fields.

$$\begin{aligned}\mathbf{H}_\mathrm{i} &= \frac{n_1}{c} - \hat{x}E_\mathrm{i}\exp j[k_\mathrm{i}(y\sin\theta_\mathrm{i} + z\cos\theta_\mathrm{i}) - \omega_\mathrm{i}t]\\ \mathbf{H}_\mathrm{r} &= \frac{n_1}{c}\hat{x}E_\mathrm{r}\exp j[k_\mathrm{r}(y\sin\theta_\mathrm{i} - z\cos\theta_\mathrm{i}) - \omega_\mathrm{r}t]\\ \mathbf{H}_\mathrm{t} &= \frac{n_2}{c}\hat{x}E_\mathrm{t}\exp j[k_t(y\sin\theta_\mathrm{t} + z\cos\theta_\mathrm{t}) - \omega_\mathrm{t}t]\end{aligned} \tag{4.62}$$

We need the $\hat{x}$ component of the magnetic field (parallel to the plane of incidence) to be continuous at $z = 0$, that is:

$$\begin{aligned}\frac{n_1}{c}(-\hat{x}E_\mathrm{i} + \hat{x}E_\mathrm{r}) &= \frac{n_2}{c}\hat{x}E_\mathrm{t}\\ n_1(E_\mathrm{i} + E_\mathrm{r}) &= n_2E_\mathrm{t}\end{aligned} \tag{4.63}$$

We will now use equations (4.61), and (4.63) to calculate the reflection and transmissions coefficients.

$$\begin{aligned}E_\mathrm{i} + E_\mathrm{r} &= \frac{\cos\theta_\mathrm{t}}{\cos\theta_\mathrm{i}}E_\mathrm{t}\\ E_\mathrm{i} - E_\mathrm{r} &= \frac{n_2}{n_1}E_\mathrm{t}\end{aligned} \tag{4.64}$$

Adding both expressions from (4.64), and solving for the ratio $E_\mathrm{t}/E_\mathrm{i}$:

$$T_\mathrm{p} = \frac{E_\mathrm{t}}{E_\mathrm{i}} = \frac{2n_1\cos\theta_\mathrm{i}}{n_1\cos\theta_\mathrm{t} + n_2\cos\theta_\mathrm{i}} \tag{4.65}$$

To obtain the ratio $E_\mathrm{r}/E_\mathrm{i}$ we will subtract the expressions from (4.64) to obtain:

$$\Gamma_\mathrm{p} = \frac{E_\mathrm{r}}{E_\mathrm{i}} = \frac{n_1\cos\theta_\mathrm{t} - n_2\cos\theta_\mathrm{i}}{n_1\cos\theta_\mathrm{t} + n_2\cos\theta_\mathrm{i}} \tag{4.66}$$

We can also express these coefficients in terms of the angle of incidence, and the refractive index of both materials by applying Snell's Law:

$$\Gamma_p = \frac{n_1\sqrt{1-\left(\frac{n_1}{n_2}\sin\theta_i\right)^2} - n_2\cos\theta_i}{n_1\sqrt{1-\left(\frac{n_1}{n_2}\sin\theta_i\right)^2} + n_2\cos\theta_i} \tag{4.67}$$

and

$$T_p = \frac{2n_1\sqrt{1-\left(\frac{n_1}{n_2}\sin\theta_i\right)^2}}{n_1\sqrt{1-\left(\frac{n_1}{n_2}\sin\theta_i\right)^2} + n_2\cos\theta_i} \tag{4.68}$$

4.3.3 Conservation of power

It is convenient to express the Fresnel coefficients in terms of their optical power because that is the parameter that we detect when using power meters. Energy conservation requires that the incident power at the interface is equal to the reflected and transmitted power, that is:

$$P_i = P_r + P_t \tag{4.69}$$

we can define the ratio of the reflected power to the incident power as *reflectance*, $\mathcal{R}$, and the ratio between the transmitted power and the incident power as *transmittance*, $\mathcal{T}$, that is:

$$\mathcal{R} = \frac{P_r}{P_i} = \frac{|E_r|^2}{|E_i|^2} = |\Gamma|^2, \tag{4.70}$$

where we are using the fact that the power can be expressed in terms of the square of their electric fields (i.e. their intensities). We can then rewrite our conservation of power equation as:

$$\mathcal{R} + \mathcal{T} = 1 \tag{4.71}$$

To calculate the transmittance value, however, we need to consider the angle of incidence and reflection and the refractive index of the materials at the interface [1, 2].

$$\mathcal{T} = \frac{n_2\cos\theta_t}{n_1\cos\theta_i}|T|^2 \tag{4.72}$$

Example 4.3. Show that $\mathcal{R} + \mathcal{T} = 1$, for an s-polarized wave.

We first calculate an expression for $\mathcal{R}$

$$\begin{aligned}\mathcal{R} &= \left| \frac{n_1 \cos\theta_i - n_2 \cos\theta_t}{n_1 \cos\theta_i + n_2 \cos\theta_t} \right|^2 \\ &= \frac{n_1^2 \cos^2\theta_i - 2n_1 n_2 \cos\theta_i \cos\theta_t + n_2^2 \cos^2\theta_t}{(n_1 \cos\theta_i + n_2 \cos\theta_t)^2}\end{aligned} \tag{4.73}$$

We now calculate $\mathcal{T}$

$$\begin{aligned}\mathcal{T} &= \frac{n_2 \cos\theta_t}{n_1 \cos\theta_i} \left| \frac{2n_1 \cos\theta_i}{n_1 \cos\theta_i + n_2 \cos\theta_t} \right|^2 \\ &= \frac{4n_1 n_2 \cos\theta_i \cos\theta_t}{(n_1 \cos\theta_i + n_2 \cos\theta_t)^2}\end{aligned} \tag{4.74}$$

we now have the following expression:

$$\begin{aligned}\mathcal{R} + \mathcal{T} &= \frac{n_1^2 \cos^2\theta_i + 2n_1 n_2 \cos\theta_i \cos\theta_t + n_2^2 \cos^2\theta_t}{(n_1 \cos\theta_i + n_2 \cos\theta_t)^2} \\ &= \frac{(n_1 \cos\theta_i + n_2 \cos\theta_t)^2}{(n_1 \cos\theta_i + n_2 \cos\theta_t)^2} = 1\end{aligned} \tag{4.75}$$

a similar exercise can be done for p-polarized waves.

4.3.4 Brewster angle

Let's take a closer look at the reflection coefficients for both polarizations. We can graph the value of Γ for an interface between two materials like air, $n_1 = 1$, and glass, $n_2 = 1.5$. This is shown in figure 4.6. A negative reflection coefficient corresponds to a phase shift of π upon reflection.

Notice that at some angle, Γ_p crosses a zero value, indicating that at this angle, there is zero p-polarized light being reflected. Also, the transition from a negative reflection coefficient to a positive reflection coefficient mean a phase change π. This particular angle is called the *Brewster angle*.

The Brewster angle is given by the formula:

$$\theta_B = \arctan\left(\frac{n_2}{n_1}\right) \tag{4.76}$$

where n_1 is the refractive index of the medium in which the light is incident, and n_2 is the refractive index of the dielectric material. The Brewster angle depends on the refractive indices of the materials and is different for different materials and different wavelengths of light.

The Brewster angle is an important concept in many areas of optics, including polarimetry, where it is used to minimize reflected light in optical systems, and in

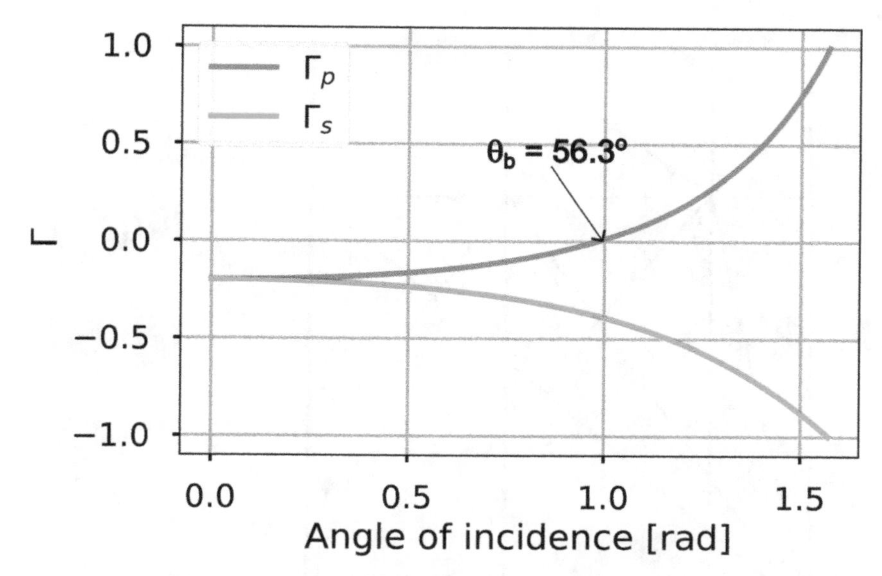

Figure 4.6. Reflection coefficients for an air–glass interface.

display technology, where it is used to reduce glare and improve the contrast of images. In chapter 13, we show a refractive index sensor based on the Brewster angle.

4.4 Evanescent waves

We will now look at the boundary conditions when we are at a critical angle between two materials, as shown in figure 4.7 Let's start by using Snell's law and calculate the angle of refraction when light travels between two different materials.

$$\theta_t = \arcsin\left(\frac{n_1}{n_2}\sin\theta_i\right) \tag{4.77}$$

we can see that if $n_1 > n_2$ there is an angle θ_i at which θ_t becomes imaginary. We call this angle *the critical angle*, defined as:

$$\theta_c = \arcsin\left(\frac{n_2}{n_1}\right) \tag{4.78}$$

and if we exceed this angle of incidence, we will experience total internal reflection, and the reflection coefficient will be equal to 1, regardless of the polarization. We can show this by plotting equations (4.56) and (4.67) for a light beam going from glass, $n_1 = 1.5$, to air, $n_2 = 1$. However, we will make a correction when we are working above the critical angle $\theta_i > \theta_c$:

$$\cos\theta_2 = j\sqrt{\left(\frac{n_1}{n_2}\sin\theta_i\right)^2 - 1} \tag{4.79}$$

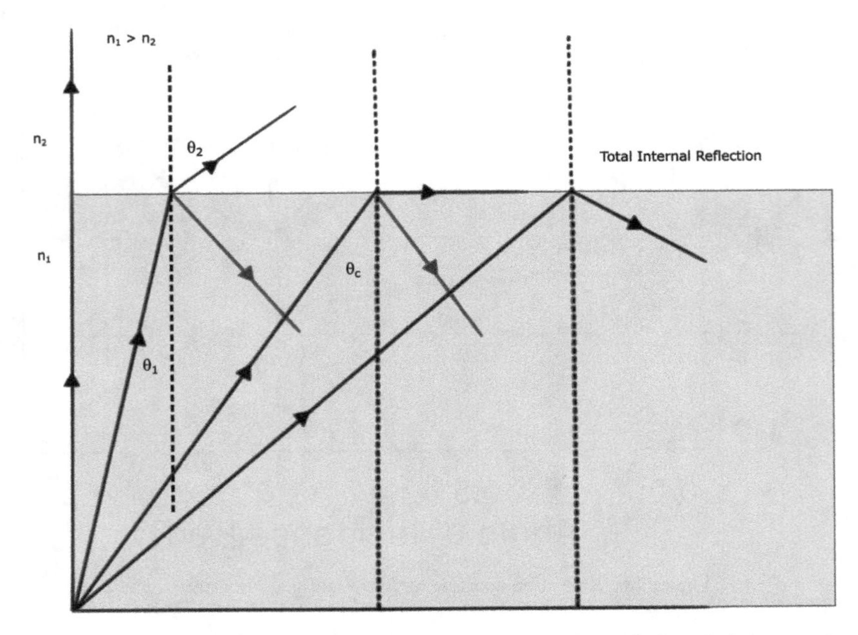

Figure 4.7. Light refraction between two different materials, above the critical angle light experiences total internal reflection. Image created by L Argueta-Slaughter.

substituting this expression into equations (4.56) and (4.67), we obtain an expression that is valid when $\theta_\mathrm{i} > \theta_\mathrm{c}$:

$$\Gamma_\mathrm{p} = \frac{jn_1\sqrt{\left(\frac{n_1}{n_2}\sin\theta_\mathrm{i}\right)^2 - 1} - n_2\cos\theta_\mathrm{i}}{jn_1\sqrt{\left(\frac{n_1}{n_2}\sin\theta_\mathrm{i}\right)^2 - 1} + n_2\cos\theta_\mathrm{i}} \tag{4.80}$$

and,

$$\Gamma_\mathrm{s} = \frac{n_1\cos\theta_\mathrm{i} - jn_2\sqrt{\left(\frac{n_1}{n_2}\sin\theta_\mathrm{i}\right)^2 - 1}}{n_1\cos\theta_\mathrm{i} + jn_2\sqrt{\left(\frac{n_1}{n_2}\sin\theta_\mathrm{i}\right)^2 - 1}}, \tag{4.81}$$

In figure 4.8 we are plotting the magnitudes of equations (4.56) and (4.67) for $\theta_\mathrm{i} < \theta_\mathrm{c}$, and equations (4.81) and (4.80) for $\theta_\mathrm{i} > \theta_\mathrm{c}$

We can see that above the critical angle, the reflection coefficient equals one, as expected. One interesting aspect of total internal reflection is when we analyze the boundary conditions at the interface. As we have done before, we need the parallel components of the electric field to be continuous at the interface, but if there is no

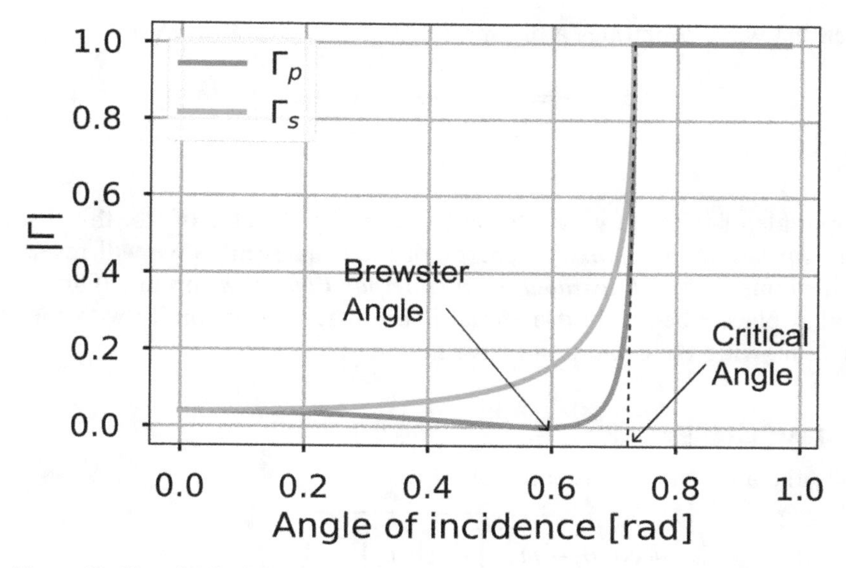

Figure 4.8. Plot of reflection coefficients for *s*- and *p*-polarization for a glass–air interface.

electric field being transmitted, it may seem that we got ourselves into a contradiction.

Let's look at an electric field on the second material using the same models as in figure 4.5:

$$\mathbf{E}_t = \mathbf{E}_{0t} \exp j[k_t(y \sin\theta_t + z \cos\theta_t) - \omega_t t] \tag{4.82}$$

where $\mathbf{E}_{0t}$ includes the *s*- and *p*-components of the transmitted electric field. Above the critical angel $\cos\theta_t$ will be imaginary, so we can rewrite it using equation (4.79), and our propagation vector, $\mathbf{k} \cdot \mathbf{r}$ can be redefine as:

$$\mathbf{k} \cdot \mathbf{r} = k_t y \frac{n_1}{n_2} \sin\theta_i + jk_t z \sqrt{\left(\frac{n_1}{n_2} \sin\theta_i\right)^2 - 1} \tag{4.83}$$

to simplify our previous equation, let's define a factor α as:

$$\alpha = k_t \sqrt{\left(\frac{n_1}{n_2} \sin\theta_i\right)^2 - 1} \tag{4.84}$$

we can now, express the transmitted electric field as:

$$\mathbf{E}_t = \mathbf{E}_{0t} \exp(-j\omega t) \exp\left(jk_t y \frac{n_1}{n_2} \sin\theta_i\right) \exp(-\alpha z) \tag{4.85}$$

The last factor on the right describes an exponential decrease of the electric field as it enters the second medium. We call α the *attenuation factor*.

When the wave penetrates a distance:

$$z = \frac{1}{\alpha} = \frac{\lambda}{2\pi\sqrt{\left(\frac{n_1}{n_2}\sin\theta_i\right)^2 - 1}} \tag{4.86}$$

the amplitude of the field would have decreased by a factor of $1/e$, this distance is known as *penetration depth*. The energy of the evanescent wave will revert to the initial medium, unless *Frustrated Total Internal Reflection*, occurs. This phenomenon takes place when a third material is brought into proximity with the second material, allowing the evanescent wave to persist.

4.5 Phase change

We will take a closer look at equation (4.81).

$$\Gamma_s = \frac{n_1\cos\theta_i - jn_2\sqrt{\left(\frac{n_1}{n_2}\sin\theta_i\right)^2 - 1}}{n_1\cos\theta_i + jn_2\sqrt{\left(\frac{n_1}{n_2}\sin\theta_i\right)^2 - 1}} = \frac{a - jb}{a + jb} \tag{4.87}$$

where $a = n_1\cos\theta_i$ and $b = n_2\sqrt{\left(\frac{n_1}{n_2}\sin\theta_i\right)^2 - 1}$

We can express Γ_s in phasor notation as:

$$\Gamma_s = \frac{\sqrt{a^2 + b^2}\quad e^{-j\Phi_s}}{\sqrt{a^2 + b^2}\quad e^{j\Phi_s}} = e^{-2j\Phi_s} \tag{4.88}$$

where

$$\tan(\Phi_s) = \left[\frac{n_2}{n_1\cos\theta_i}\sqrt{\left(\frac{n_1}{n_2}\sin\theta_i\right)^2 - 1}\right] \tag{4.89}$$

similarly, for Γ_p, we obtain a phase shift of:

$$\tan(\Phi_p) = \left[\frac{n_1}{n_2\cos\theta_t}\sqrt{\left(\frac{n_1}{n_2}\sin\theta_i\right)^2 - 1}\right] \tag{4.90}$$

Figure 4.9 shows the values of Φ obtain from equations (4.89) and (4.90). An observation that stands out is that for p-polarized waves, we have a phase shift before and after the Brewster angle ($0 \rightarrow \pi$), and once we reach the critical angle, the phase will change according to equation (4.90). Similarly, for an s-polarized wave, the phase shift remains constant and once we reach the critical angle, it will change according to equation (4.89).

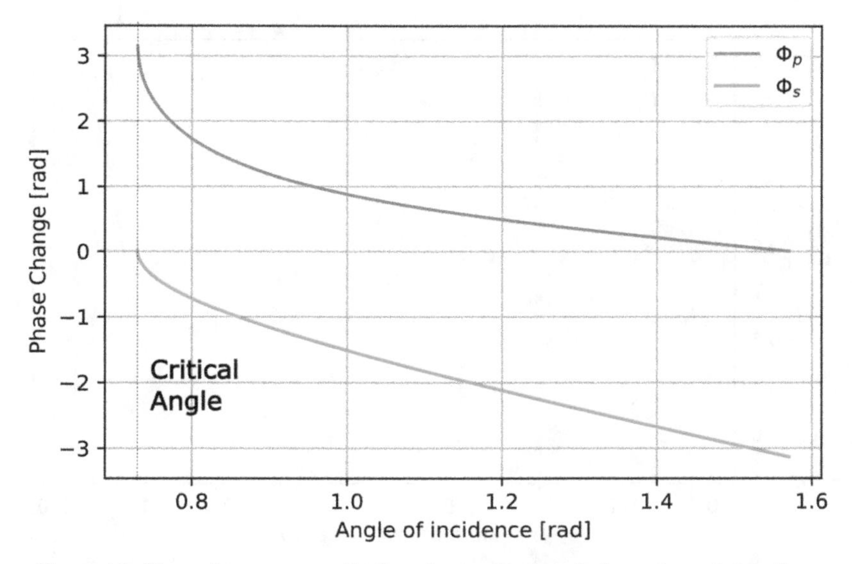

Figure 4.9. Phase change upon reflection above critical angle for a glass–air interface.

4.6 Reflection on a metallic interface

In this final section, we will look at the interaction of light with a metallic surface, or in general, with materials with a complex refractive index. For a propagating wave, the real part of the complex refractive index is a measure of the degree to which light is slowed down as it enters the material, compared to its speed in a vacuum. In contrast, the imaginary part represents the attenuation of the wave as it propagates in such materials. A material with a high imaginary component of the complex refractive index will absorb more light than a material with a low imaginary component. We can define our complex refractive index as follows:

$$\tilde{n} = n + j\kappa \tag{4.91}$$

One of the advantages of Fresnel's coefficients is that they are valid regardless of whether the refractive index is complex or not. We will just have to substitute $\tilde{n}$ instead of n. We can rewrite equations (4.56) and (4.67) as:

$$\Gamma_\mathrm{p} = \frac{n_1\sqrt{1 - \left(\frac{n_1}{\tilde{n}}\sin\theta_\mathrm{i}\right)^2} - \tilde{n}\cos\theta_i}{n_1\sqrt{1 - \left(\frac{n_1}{\tilde{n}}\sin\theta_\mathrm{i}\right)^2} + \tilde{n}\cos\theta_i} \tag{4.92}$$

and,

$$\Gamma_\mathrm{s} = \frac{n_1\cos\theta_\mathrm{i} - \tilde{n}\sqrt{1 - \left(\frac{n_1}{\tilde{n}}\sin\theta_\mathrm{i}\right)^2}}{n_1\cos\theta_\mathrm{i} + \tilde{n}\sqrt{1 - \left(\frac{n_1}{\tilde{n}}\sin\theta_\mathrm{i}\right)^2}}, \tag{4.93}$$

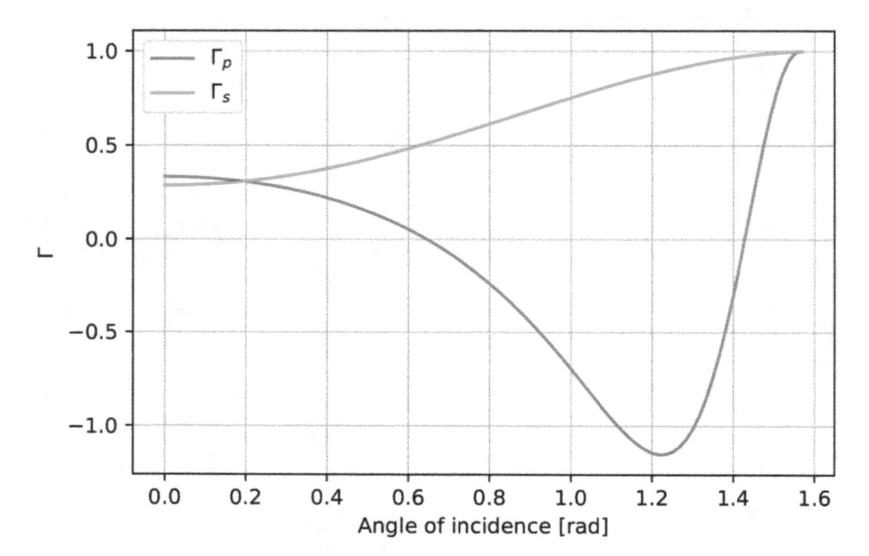

Figure 4.10. Reflection coefficients for a glass–gold interface.

We can now follow a procedure similar to the one used in section 4.5 to obtain an expression for the Fresnel coefficients for a metallic interface. As an example, figure 4.10 shows the Fresnel coefficients for an interface between glass and gold $(\tilde{n}) = 0.181 + 3.07j$

References

[1] Born M *et al* 1999 *Principles of Optics* 7th edn (Cambridge: Cambridge University Press)

[2] Saleh B E A and Teich M C 2019 *Fundamentals of Photonics* (New York: Wiley)

IOP Publishing

Optical Sensors
An introduction with lab demonstrations
Victor Argueta-Diaz

Chapter 5

Physical optics

5.1 Introduction

In the last two chapters, we established the basis for using Maxwell equations to describe electromagnetic waves. We discussed electromagnetic wave propagation and looked at how it behaves at the interface of different materials. We will now continue our study of waves and look at different phenomena like diffraction and interference. This particular area of study is usually referred as: *physical optics*.

Physical optics is a branch of optics that deals with the wave-like properties of light, including diffraction, interference, polarization, and other phenomena that the ray model of light cannot explain. In physical optics, light is treated as an electromagnetic wave, which can be described using Maxwell's equations. This approach is used to describe phenomena such as light scattering, diffraction of light by gratings and lenses, interference, and the behavior of optical instruments.

Physical optics makes use of the theory of wave optics, which is based on a mathematical description of the propagation of light in terms of its electric and magnetic fields. Wave optics also uses the Huygens–Fresnel principle to describe how light propagates and interacts with objects. It also uses the boundary conditions of Maxwell's equations to calculate the reflection and refraction of light at the surfaces of materials. By combining these theories, physical optics can be used to describe the properties of light and how it interacts with matter.

This chapter describes the phenomenon of interference, which refers to the phenomenon that occurs when two or more light waves interact with each other, resulting in a modification of the amplitude, phase, or polarization of the light. This interaction can either enhance or cancel out certain regions of the light wave, depending on the relative phase difference between the waves. In the next chapter, we will be looking at the phenomenon of diffraction. Diffraction is a property of waves that describes how they alter their direction as they pass through a narrow opening.

doi:10.1088/978-0-7503-4876-8ch5

5.2 Optical interference

Before describing optical interference, we need to talk about superposition of waves. Let's assume that, Ψ_1, and Ψ_2 are two independent solutions of our wave equation such that:

$$\nabla^2\Psi_i + \omega^2\mu\varepsilon\Psi_i = 0 \tag{5.1}$$

then any linear combination of both solutions is also a solution of the wave equations.

$$\Psi = C_1\Psi_1 + C_2\Psi_2 \tag{5.2}$$

where C_1, and C_2 are real constants.

Let's consider two propagating plane waves in space, that overlap at a point in space; we will assume that the electric fields of each wave can be described by the following expressions:

$$\mathbf{E}_1 = E_1 \cos(\mathbf{k}_1 \cdot \mathbf{r}_1 - \omega t + \phi_1) \tag{5.3}$$

$$\mathbf{E}_2 = E_2 \cos(\mathbf{k}_2 \cdot \mathbf{r}_2 - \omega t + \phi_2) \tag{5.4}$$

where ω_i, k_i, and ϕ_i, are the angular frequency, propagation constant, and phase for each wave, respectively. At the point where both waves overlap, we can calculate the total amplitude of the wave will be:

$$\mathbf{E} = \mathbf{E}_2 + \mathbf{E}_2 \tag{5.5}$$

and its irradiance will be given by:

$$\begin{aligned} I &= \langle \mathbf{E} \cdot \mathbf{E}^* \rangle \\ &= \langle (\mathbf{E}_1 + \mathbf{E}_2) \cdot (\mathbf{E}_1 + \mathbf{E}_2) \rangle \\ &= \langle \mathbf{E}_1^2 + \mathbf{E}_2^2 + 2\mathbf{E}_1 \cdot \mathbf{E}_2 \rangle \end{aligned} \tag{5.6}$$

where $\langle \cdots \rangle$ represents the temporal average. Equation (5.6) is the sum of the intensities of our two fields, plus a cross-factor.

$$I = I_1 + I_2 + I_{12} \tag{5.7}$$

The first two terms, are just $I_1 = \frac{E_1^2}{2}$, and $I_2 = \frac{E_2^2}{2}$. The last term, $I_{12} = 2\langle \mathbf{E}_1 \cdot \mathbf{E}_2 \rangle$, is a little bit more complicated. It is called the *interference term*, and taking a closer look at it, we have the following:

$$\mathbf{E}_1 \cdot \mathbf{E}_2 = E_1 E_2 \cos(\mathbf{k}_1 \cdot \mathbf{r}_1 - \omega t + \phi_1)\cos(\mathbf{k}_2 \cdot \mathbf{r}_2 - \omega t + \phi_2) \tag{5.8}$$

using the identity, $\cos(a - b) = \cos(a)\cos(b) + \sin(a)\sin(b)$, and letting $a = \mathbf{k} \cdot \mathbf{r} + \phi$, and $b = \omega t$, we can rewrite our last expression as:

$$\begin{aligned} \mathbf{E}_1 \cdot \mathbf{E}_2 = E_1 E_2 &\left[\cos(\mathbf{k}_1 \cdot \mathbf{r}_1 + \phi_1)\cos(\omega t) + \sin(\mathbf{k}_1 \cdot \mathbf{r}_1 + \phi_1)\sin(\omega t)\right] \\ &\left[\cos(\mathbf{k}_2 \cdot \mathbf{r}_2 + \phi_2)\cos(\omega t) + \sin(\mathbf{k}_2 \cdot \mathbf{r}_2 + \phi_2)\sin(\omega t)\right] \end{aligned} \tag{5.9}$$

and taking the time average of equation (5.9) we obtain [1, 2]:

$$\langle \mathbf{E}_1 \cdot \mathbf{E}_2 \rangle = \frac{1}{2} E_1 E_2 \cos(\mathbf{k}_1 \cdot \mathbf{r}_1 + \phi_1 - \mathbf{k}_2 \cdot \mathbf{r}_2 - \phi_2) \tag{5.10}$$

We can then express the electric field of two waves overlapping at a point in space as:

$$\langle \mathbf{E}_1^2 + \mathbf{E}_2^2 + 2\mathbf{E}_1 \cdot \mathbf{E}_2 \rangle = \frac{E_1^2}{2} + \frac{E_1^2}{2} + E_1 E_2 \cos(\mathbf{k}_1 \cdot \mathbf{r}_1 + \phi_1 - \mathbf{k}_2 \cdot \mathbf{r}_2 - \phi_2) \tag{5.11}$$

although it is more common to express this equation in terms of intensities as:

$$I = I_1 + I_2 + 2\sqrt{I_1 I_2} \cos(\delta) \tag{5.12}$$

where δ is the phase difference and its defined as:

$$\delta = \mathbf{k}_1 \cdot \mathbf{r}_1 + \phi_1 - \mathbf{k}_2 \cdot \mathbf{r}_2 - \phi_2 \tag{5.13}$$

We can see that depending on the value of δ, the total irradiance presents two extreme values. The first extreme is when $\delta = 2m\pi$, where m is any integer value or zero, the irradiance will be:

$$I_{\max} = I_1 + I_2 + 2\sqrt{I_1} I_2 \tag{5.14}$$

We call this case *constructive interference.* The second extreme, happens when $\delta = (2\,m + 1)\pi$, in which case $\cos(\delta) = -1$, and our irradiance will be

$$I_{\min} = I_1 + I_2 - 2\sqrt{I_1} I_2 \tag{5.15}$$

which is the case for *destructive interference*. If we have two sources with the same amplitude we can simplify equations (5.12) to:

$$I = 2I_0(1 + \cos(\delta)) = 4I_0 \cos^2\left(\frac{\delta}{2}\right) \tag{5.16}$$

Let's assume two monochromatic waves with equal amplitude traveling from source S_1 and S_2 towards an overlapping point P as shown in figure 5.1 in this case, our irradiance equations will be:

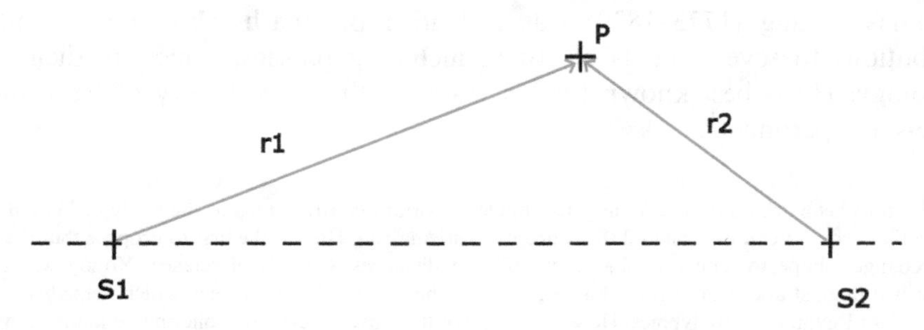

Figure 5.1. Waves traveling from different points in space and overlapping at point P.

$$I = 2I_0\left(1 + \cos\left[k(r_1 - r_2) + (\phi_1 - \phi_2)\right]\right) \tag{5.17}$$

Our maximum values will occur when:

$$k(r_1 - r_2) + (\phi_1 - \phi_2) = 2\pi m \tag{5.18}$$

in terms of $(r_1 - r_2)$:

$$(r_1 - r_2) = \frac{2\pi m - (\phi_1 - \phi_2)}{k} \tag{5.19}$$

if we assume that the sources are in phase such that $\phi_1 - \phi_2 = 0$, then we have a condition for constructive interference:

$$(r_1 - r_2) = \frac{2\pi m}{k} = \lambda m \tag{5.20}$$

we can repeat a similar procedure for destructive interference, by replacing m by $(2\,m + 1)$ to obtain the conditions for destructive interference:

$$(r_1 - r_2) = \frac{2\pi(2\,m + 1)}{k} = \lambda(2\,m + 1) \tag{5.21}$$

At different points away from S_1, and S_2, the conditions for constructive and destructive interference will be met. Creating a pattern of bright and dark fringes. This pattern is called an *interference pattern*. We can observe that there will be some planes (i.e. columns) in which we will have constructive interference for every point on that plane. In particular, when $m = 0$, it will describe a plane equidistant to both points where we will have constructive interference. From that central fringe, we will alternate between dark and bright fringes to the left and right. Each bright fringe will represent a new value of m that will satisfy our constructive interference conditions. Figure 5.2 shows the interference pattern for two plane waves.

5.2.1 Double slit

Young's double-slit experiment is a classic experiment in physics that demonstrates the wave nature of light. It was first performed by the British physicist Thomas Young in 1801.

Thomas Young (1773–1829) was a British polymath who made significant contributions to several fields of study, including physics, optics, medicine, and Egyptology. He is best known for his work on the wave theory of light and his double-slit experiment[1].

[1] In addition to his work on optics, Young also made important contributions to the study of human vision, including the theory of color vision and the concept of astigmatism. He was the first to propose that the lens of the eye changes shape to focus on objects at different distances. Outside of science, Young was also an accomplished linguist and is known for his work in deciphering the Rosetta Stone, which helped scholars to decode ancient Egyptian hieroglyphics. He was a fellow of the Royal Society of London and made many other contributions to various fields during his lifetime [3].

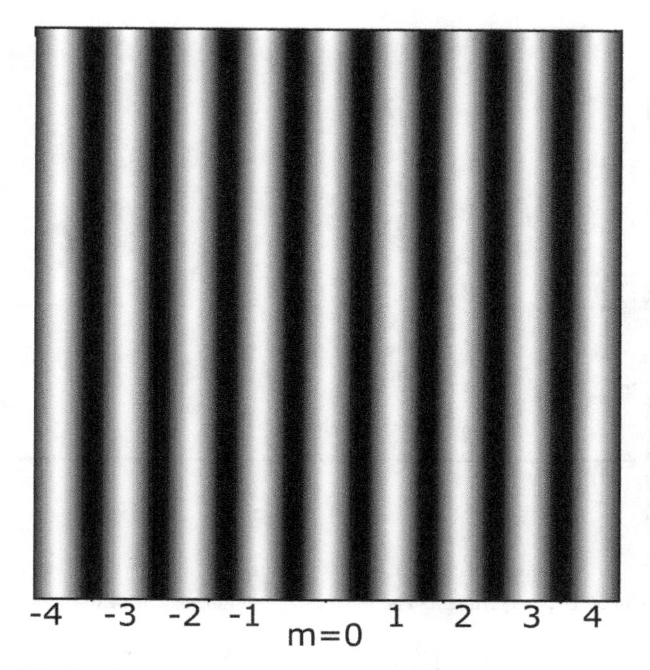

Figure 5.2. Interference pattern for two plane waves. Computer-generated image.

The double-slit experiment involves shining a monochromatic light source, such as a laser, at a screen with two narrow slits in it. The light passing through the two slits diffracts and creates an interference pattern of light and dark fringes on a second screen placed behind the first. Our task then, is to understand the conditions of constructive and destructive interference and describe the interference pattern. Figure 5.3 shows the schematic of the double-slit experiment.

In Young's experiment, we see two slits that function as light sources, S_1 and S_2. The slits are separated a distance d, of a few millimeters, and the aperture of each slit is of a couple of λs; these slits can be illuminated with a stable monochromatic light source, (e.g. a laser, not shown in the picture) to guarantee same amplitude and phase difference between sources. Light from each slit will propagate a distance L to a viewing screen, we will assume that $L \gg d$. We will examine the constructive interference condition at any point P on the viewing screen.

In order to have constructive interference at an arbitrary point P we need to satisfy the condition set in equation (5.20). From our schematic we can see that point P, is located at a distance y from the optical axis. This distance can be found from basic geometry as:

$$y = L\tan(\theta) \tag{5.22}$$

where θ is the angle between an equidistant point between the two slits, the optical axis, and our point P. Because of our condition that $L \gg d$, we can say that the two paths to point are almost parallel to each other, so the small shaded triangle in the

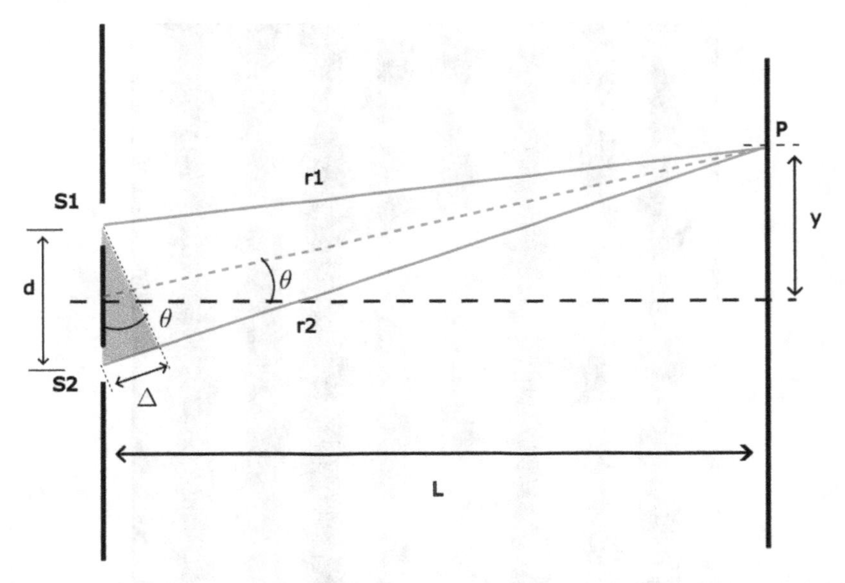

Figure 5.3. Young's double-slit experiment. The separation and dimensions of the slits is comparable to the light's wavelength.

figure can be a triangle rectangle, with also an angle θ. The path difference between r_1, and r_2 is then given by the length of the short side of the shaded triangle.

$$\Delta r = d\sin(\theta) \tag{5.23}$$

Our bright fringes will occur at different angles θ_m, such that:

$$\Delta r = d\sin(\theta_m) = m\lambda \tag{5.24}$$

Due to the dimensions of the slits, and their spacing, we can assume that θ_m will be very small, and we can use the small-angle approximation and simplify the previous equations to:

$$\theta_m = \frac{m\lambda}{d} \tag{5.25}$$

and our position y can be simplified, again using a small-angle approximation, to:

$$y = \frac{m\lambda L}{d} \tag{5.26}$$

In terms of our path difference, Δr, we can rewrite equation (5.26) as,

$$\Delta r = m\lambda = \frac{yd}{L} \tag{5.27}$$

From equation (5.17), we can express δ in terms the y-position of our point P as:

$$\delta = k\Delta r \tag{5.28}$$

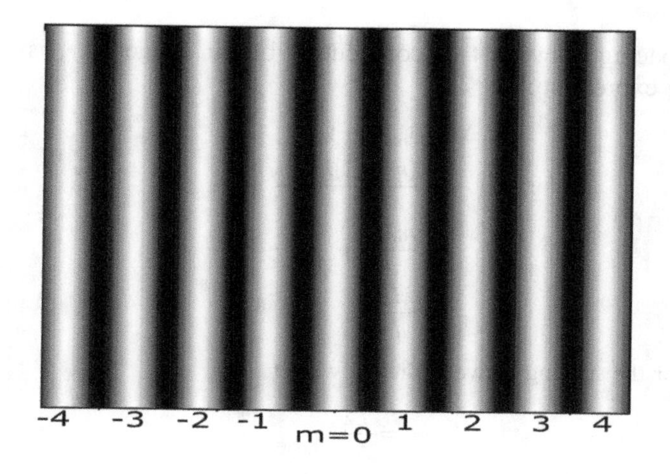

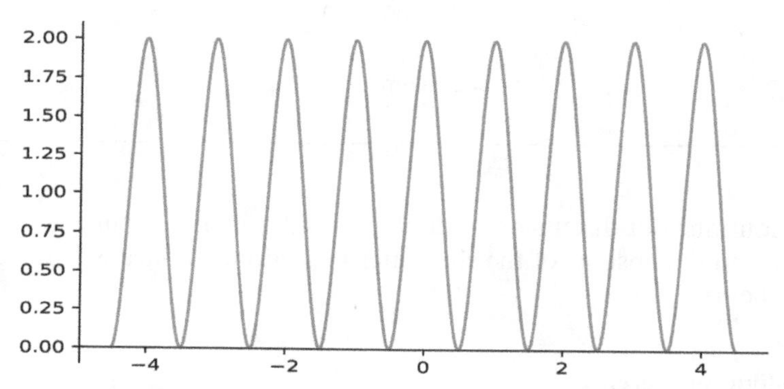

Figure 5.4. Simulation of bright fringes on an interference pattern corresponding to integer values of m.

$$= \frac{2\pi}{\lambda} \frac{yd}{L} \tag{5.29}$$

and the irradiance, from equation (5.16), as:

$$I = 4I_0 \cos^2\left(\frac{\pi yd}{\lambda L}\right) \tag{5.30}$$

We can see the plot of this equation in figure 5.4, alongside an interference pattern created by two plane waves. We can see that the maximum values correspond to bright fringes and integer values of m.

Example 5.1. Light from a monochromatic source with $\lambda = 550$ nm illuminates two narrow slits. The distance between two consecutive fringes on a screen 2 meters away is 5 mm. What's the spacing between the slits?

The spacing between two consecutive bright fringes, Δy, is given by the following expression:

$$\begin{aligned}\Delta y &= y_{m+1} - y_m \\ &= \frac{(m+1)\lambda L}{d} - \frac{m\lambda L}{d} \\ &= \frac{m\lambda L}{d} + \frac{\lambda L}{d} - \frac{m\lambda L}{d} \\ &= \frac{\lambda L}{d}\end{aligned} \tag{5.31}$$

Solving for the spacing between slits, d, we get:

$$\begin{aligned}d &= \frac{L\lambda}{\Delta y} \\ &= \frac{2 \cdot 500 \times 10^{-9}}{4 \times 10^{-3}} \\ &= 250\ \mu\text{m}\end{aligned}$$

We will continue our discussion of double-slit interference, to understand the role of the shape and dimensions of the slits, later in section 6.4 once we introduce some diffraction theory.

5.2.2 Thin-film interference

Thin-film interference is a phenomenon that occurs when light waves reflect off the upper and lower surfaces of a thin film surrounded by different materials, resulting in the interference of the reflected waves. This interference causes certain wavelengths of light to either reinforce or cancel each other out, resulting in an interference pattern.

The thickness of the film, the refractive indices of the film and the surrounding media, and the angle of incidence of the light all play a role in determining the spacing and pattern of the interference fringes.

Thin-film interference is responsible for many natural and man-made optical phenomena, such as the colors seen in soap bubbles, oil slicks, and certain types of gemstones. It is also used in many technological applications, including anti-reflective coatings on eyeglasses and camera lenses.

Assume a dielectric film of thickness d, on top of a substrate, as shown in figure 5.5. When light is incident on the film, it will be split into two beams, one reflected and one transmitted, as was discussed already in section 4.3. The transmitted beam will be again split at the interface between the film and the substrate[2].

[2] The splitting of light at each interface occurs several times, but we will only consider the first two reflected beams.

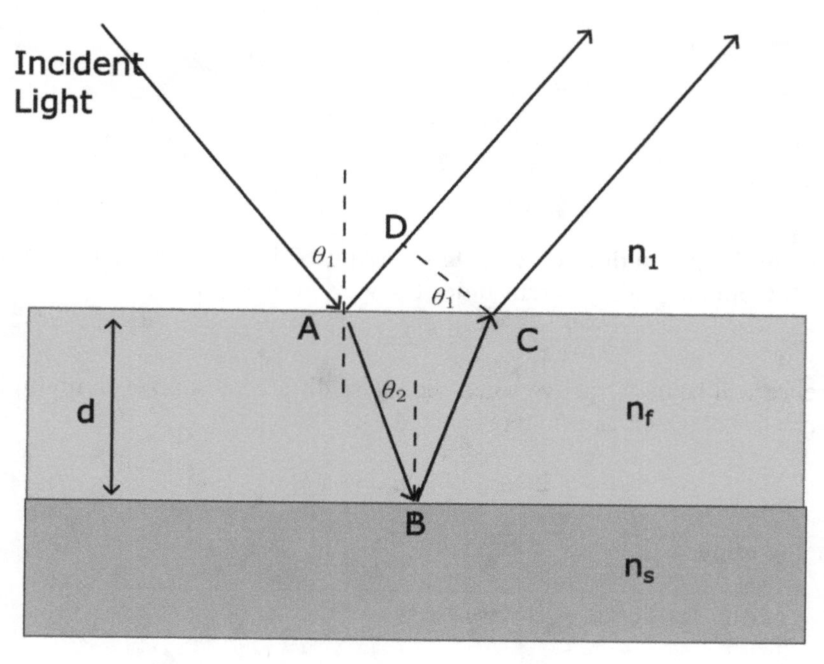

Figure 5.5. Light path difference in a thin-film structure.

The two reflected beams will be parallel after leaving the film and can be brought together at a point P, by a simple lens.

The optical path difference between the two beams will be given by:

$$\Delta = n_{\mathrm{f}}(\overline{AB} + \overline{BC}) - n_{\mathrm{l}}(\overline{AD}) \tag{5.32}$$

From geometry, we can express the segments $\overline{AB}$, and $\overline{BC}$ in terms of the thin-film thickness as:

$$\overline{AB} = \overline{BC} = \frac{d}{\cos(\theta_2)} \tag{5.33}$$

and similarly for the segment $\overline{AD}$ we have:

$$\begin{aligned} \overline{AD} &= \overline{AC}\sin(\theta_1) \\ &= \overline{AC}\frac{n_{\mathrm{f}}}{n_{\mathrm{i}}}\sin(\theta_2) \end{aligned} \tag{5.34}$$

that we obtain from Snell's law. Finally, we can write $\overline{AC}$ as:

$$\overline{AC} = 2d\tan(\theta_2) \tag{5.35}$$

Replacing equations (5.35), (5.34), and equation (5.33), into equation (5.32) get:

$$\begin{aligned}\Delta &= \frac{2n_f d}{\cos(\theta_2)} - 2n_f d \tan(\theta_2)\sin(\theta_2) \\ &= 2n_f d\left(\frac{1 - \sin^2(\theta_2)}{\cos(\theta_2)}\right) \\ &= 2n_f d \cos(\theta_2)\end{aligned} \tag{5.36}$$

To calculate the phase difference δ, that is required in equation (5.12) we need to multiply the optical path different times the wavenumber, that is:

$$\delta = k_0\Delta \tag{5.37}$$

Interference will be constructive when the phase difference is an even multiple of π, that is, $\delta = 2m\pi$, which can be reorganized as:

$$2\,m\pi = \frac{2\pi}{\lambda}2n_f d \cos(\theta_2)$$

rearranging terms:

$$d\cos(\theta_2) = \frac{m\lambda}{2n_f} \tag{5.38}$$

To find the expression for destructive interference, we need to evaluate when the phase difference is an odd multiple of π, that is, $\delta = (2m + 1)\pi$, which yields:

$$d\cos(\theta_2) = (2m + 1)\frac{\lambda}{4n_f} \tag{5.39}$$

There is, however, an additional point to make. When light is traveling from one material to a material with a higher refractive index (e.g. air to glass), there is an additional phase shift equivalent to 0.5λ, that needs to be included in our equations. Once we have included this term, our expressions for constructive and destructive interference are:

$$d\cos(\theta_2) = \begin{matrix} (2m + 1)\dfrac{\lambda}{4n_f}, & \text{constructive} \\ \dfrac{m\lambda}{2n_f}, & \text{destructive} \end{matrix} \tag{5.40}$$

This additional phase shift doesn't need to be included when light travels into a material with a lower refractive index.

Example 5.2. An drop of oil ($n_f = 1.25$), floats on water ($n_s = 1.33$). What is the minimum thickness that produces constructive interference at 500 nm. Assume an incident angle of 30°. For the same thickness, what wavelength will experience constructive interference at 70°.

The first thing we need to figure out is how many phase changes we need to account for. Light goes from air ($n_l = 1$) to oil ($n_f = 1.25$), which implies a phase change of 0.5λ, we have a second phase change, of same magnitude, when light bounces from oil into the water.

Our conditions for constructive interference will be given by equation (5.38), and solving for the film thickness:

$$d = \frac{m\lambda}{2n_f \cos(\theta_2)}$$

To calculate θ_2, we will use Snell's law.

$$n_l \sin(\theta_1) = n_f \sin(\theta_2)$$
$$\sin(\theta_2) = \frac{n_l}{n_f} \sin(\theta_1) = 0.4$$

which give us an angle $\theta_2 = 23.58°$. Substituting values, we get a minimum thickness ($m = 1$), of $d = 218.22$ nm.

We can then substitute this thickness into our previous equation, and solve instead for λ:

$$\lambda = \frac{d 2n_f \cos(\theta_2)}{m}$$

For our new angle of incidence, $\theta_1 = 70$, we get an angle of refraction of $\theta_2 = 48.74°$. This gives us a new wavelength for constructive interference of:

$$\lambda = (218.22)(2 \cdot 1.25)\cos(48.74)$$
$$= 545.55 \text{ nm}$$

In figure 5.6 we show the changes on the wavelength that exhibits constructive interference with respect of the angle of incidence. We can see how the color of the oil drop would change from violet to green, depending on the viewing angle.

5.3 Optical interferometers

An interferometer is an instrument that is used to measure and analyze the interference patterns created by the superposition of two or more waves. The basic principle behind an interferometer is to split a beam of light or other electromagnetic radiation into two separate beams, which are then recombined to create an interference pattern that can be used to measure properties of the original wave, such as its wavelength, frequency, or polarization.

There are many different types of interferometers, but they all rely on the same basic concept. Typically, an interferometer will use a beam splitter, such as a partially silvered mirror, to divide a single beam of light into two separate beams. These beams will then travel different paths, often involving mirrors or other optical elements, before being recombined at another beam splitter.

As the two beams recombine, they will create an interference pattern that depends on the phase difference between the two waves. By measuring the characteristics of

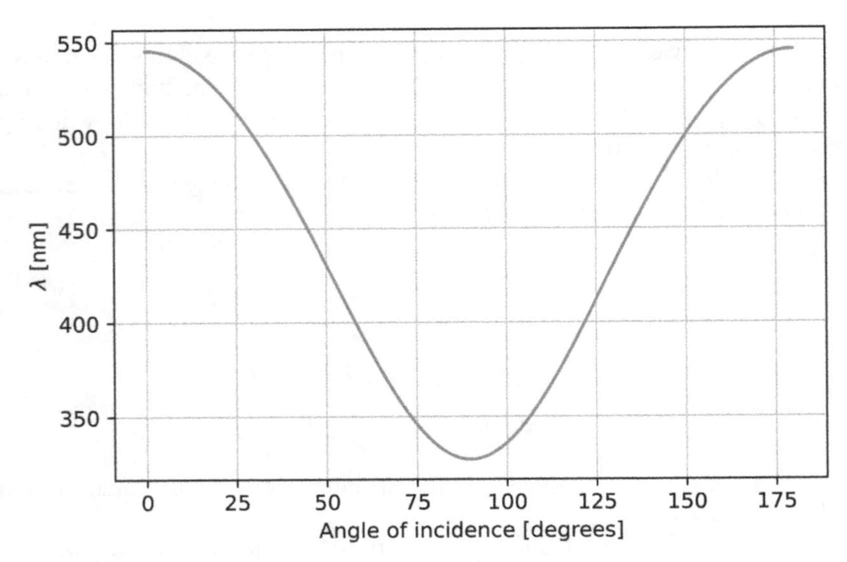

Figure 5.6. Wavelength that experiences constructive interference as a function of the incident angle.

this interference pattern, such as its intensity, shape, or fringe spacing, it is possible to determine various properties of the original wave. In this section we will describe some common interferometers.

5.3.1 Michelson interferometer

The Michelson interferometer is a type of interferometer that uses a beam splitter to split a single beam of light into two separate beams that travel perpendicular paths before being recombined at the same beam splitter. It was invented by the American physicist Albert A Michelson in the late 19th century and was used in his famous experiment to measure the speed of light.

The basic setup of a Michelson interferometer, shown in figure 5.7, consists of a light source, a beam splitter, two mirrors, and a detector. The light source sends a beam of light to the beam splitter, which splits the beam into two perpendicular paths. One path, known as the 'reference' path, goes straight to a mirror and reflects back to the beam splitter. The other path, known as the 'sample' path, goes to a second mirror before reflecting back to the beam splitter.

When the two beams are recombined at the beam splitter, they create an interference pattern that can be detected by a detector, such as a photodiode or a camera. The interference pattern depends on the path length difference between the two beams, which can be adjusted by changing the position of the mirrors. By measuring the interference pattern and analyzing its properties, it is possible to determine various properties of the light source or the objects that are being studied, such as their wavelength, frequency, or distance.

The optical path length difference, Δ, between these two beams at the detector plane is:

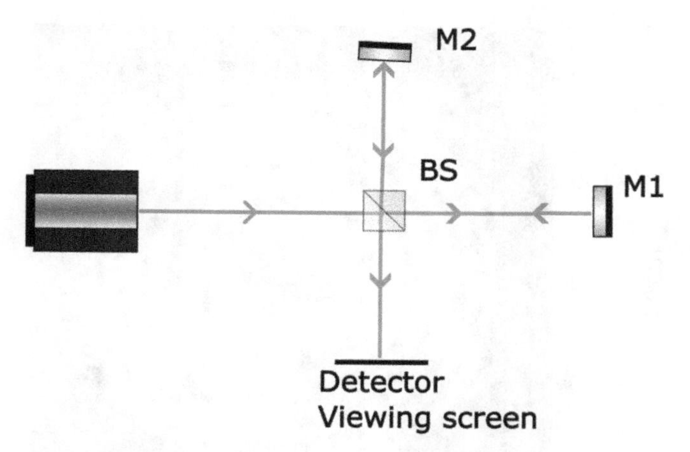

Figure 5.7. Basic setup of a Michelson Interferometer. Image designed using ComponentLibrary, created by Alexander Franzen, licensed under a Creative Commons Attribution-Non Commercial 3.0 Unported License. http://www.gwoptics.org/ComponentLibrary/.

$$\Delta = 2d\cos(\theta) \tag{5.41}$$

where d, is the relative distance between the two mirrors, and θ is the angle of the beam relative to the optical axis. The optical path difference at the center of the viewing screen is just $2d$, which corresponds to an angle $\theta = 0$. As we move our observation point away from the center, our angle θ will increase.

Our condition for constructive interference will be $\Delta = (2m + 1)\lambda$ due to the number of reflections on the beam splitter; If a condition for destructive interference is satisfied at a point P on the viewing screen, that same condition will also be satisfied for all points in a circle that are at the same distance P from the center of the viewing screen. We will expect then to see a circular interference pattern such as the one shown in figure 5.8.

The order of the central dark fringe is given by the following expression:

$$m = \frac{2d}{\lambda} \tag{5.42}$$

If we assume a $d = 1$ cm, and a wavelength $\lambda = 500$ nm, we get a mode order of $m = 40\,000$, as we move away from the center, the mode of the fringes will decrease, as θ increases. m tends to be such a large number, such that it is helpful to define a new order number p, defined as:

$$p = \frac{2d}{\lambda} - m \tag{5.43}$$

and the pth fringe is given by combining equations (5.42) and (5.43):

$$p\lambda = 2d(1 - \cos(\theta_p)) \tag{5.44}$$

The central fringe is now the $p = 0$ order, which will increase as we move away from the center.

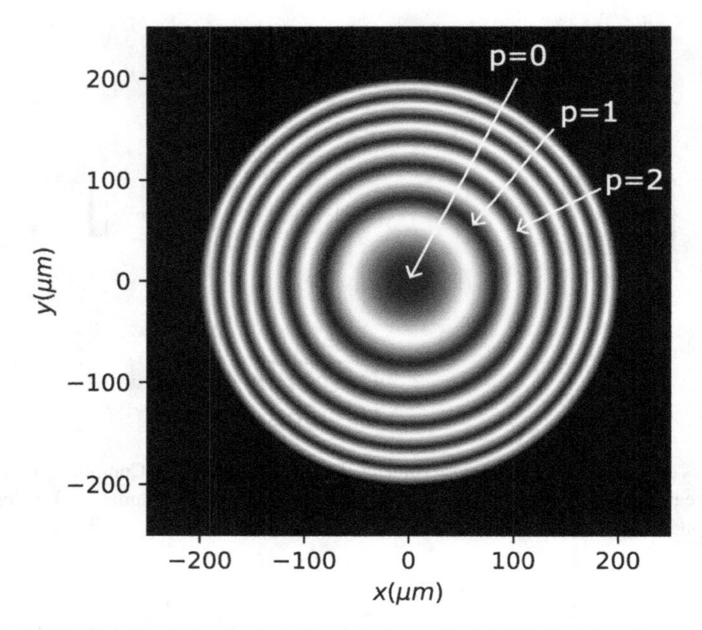

Figure 5.8. Simulated circular interference pattern for a Michelson interferometer.

One of the advantages of the Michelson interferometer, is that we can make sub-wavelength measurements. When we move a mirror a distance $\lambda/2$, then each fringe will move to the adjacent position. So by looking at point P on the viewing screen, and counting how many times there is a switch between bright and dark fringes (N) is possible to measure the mirror displacement by:

$$\Delta d = N\left(\frac{\lambda}{2}\right) \tag{5.45}$$

It is possible for the Michelson interferometer to produce an interference pattern other than concentric circles. For example, if mirrors $M1$, and $M2$ are not completely perpendicular, there will be a small angle with respect of each other. This angle difference will create a pattern of straight parallel fringes called *Fizeau fringes*; these kinds of fringes are shown in figure 5.9.

5.3.2 Mach–Zehnder interferometer

A Mach–Zehnder interferometer is an optical device that uses interference to measure small changes in the phase of light. A basic configuration is shown in figure 5.10. Light coming from a coherent light source is divided by a beam splitter (BS1). One beam travels through a reference arm, and after bouncing off a mirror (M2) it hits a second beam splitter (BS2). The second arm is the test arm. The optical path of the test arm can be changed, by changing the optical path length, the refractive index of the path, or adding testing cells. The test beam is then recombined

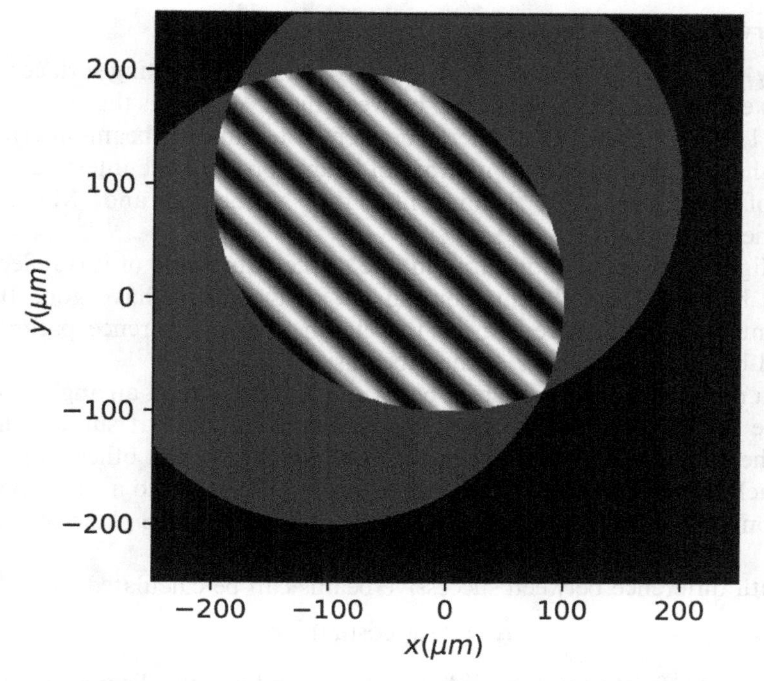

Figure 5.9. Simulated Fizeau fringes in a Michelson interferometer.

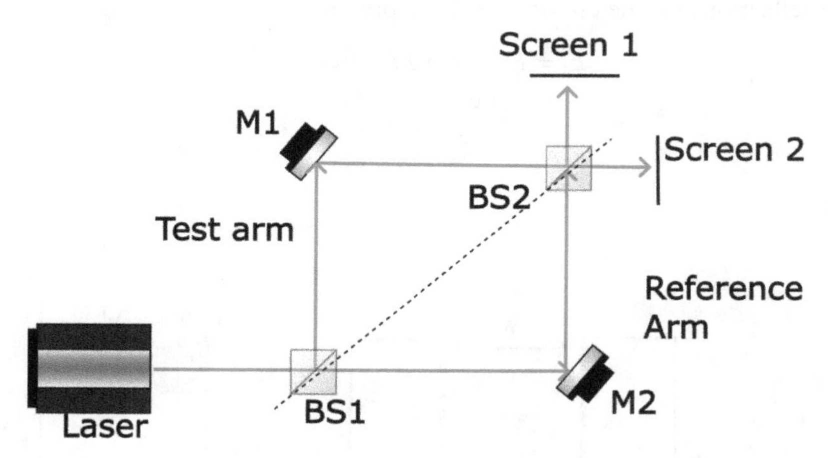

Figure 5.10. Basic configuration of a Mach–Zehnder interferometer. Image designed using Component Library, created by Alexander Franzen, licensed under a Creative Commons Attribution-Non Commercial 3.0 Unported License. http://www.gwoptics.org/ComponentLibrary/.

with the reference beam at (BS2), forming an interference pattern on two possible planes, where we can place a camera, photodetector, or viewing screen. The Mach–Zehnder interferometer is considered a variation of the Michelson interferometer, so all our derivation that we have done previously, still applies for this configuration.

5.3.3 Fabry–Perot interferometer

The Fabry–Perot interferometer consists of two partially reflecting surfaces arranged parallel to each other with a small gap between them, known as the *etalon*, shown in figure 5.11. The Fabry–Perot interferometer is a multiple-beam interferometer, meaning that an extremely high number of beams produce the interference pattern instead of just two light beams like the Michelson and Mach–Zehnder interferometers.

When light enters the etalon through one of the flats, some of it is reflected back and forth between them. Beams coming out of the etalon have gone through a different number of 'bounces' inside the etalon. The interference pattern will be calculated by adding each one of these beams.

When a monochromatic, coherent beam enters the etalon at an angle α, the beam will create multiple coherent beams as it bounces off the flat surfaces inside the etalon. The beams exiting the etalon, will be parallel to each other with an angle, also α. Each of these parallel beams can be brought together to a single point P, by using a converging lens, creating an interference pattern at the plane of the viewing screen.

The path difference between successive beams can be calculated as:

$$\Delta = 2n_e w \cos(\alpha)) \tag{5.46}$$

where n_e, is the refractive index inside the etalon, and w is the distance between flats. Let's assume that the first beam coming out of the etalon, which doesn't have any internal reflections on the etalon, can be expressed as:

$$\begin{aligned} E_1 &= E_i\sqrt{T}e^{j\delta_t}\sqrt{T}e^{-j\delta_t}e^{j\delta} \\ &= E_iTe^{i\delta} \end{aligned} \tag{5.47}$$

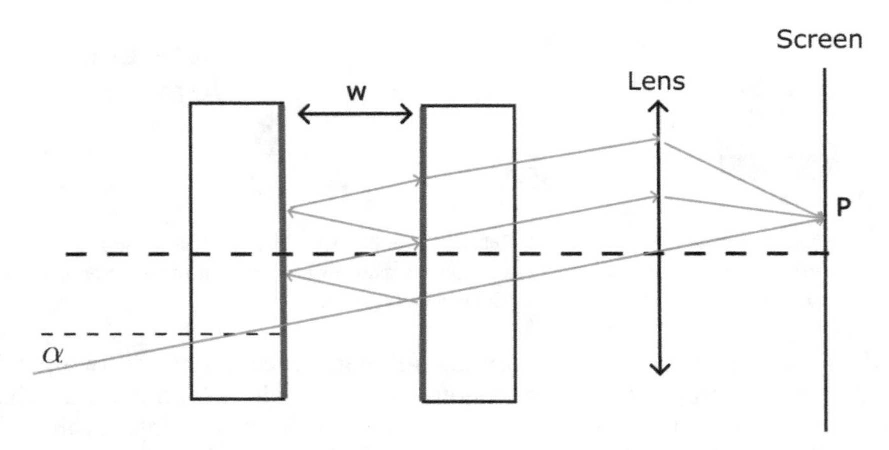

Figure 5.11. Etalon on a Fabry–Perot interferometer.

where δ_t is the phase difference of the wave traveling inside the glass, T is the transmission coefficient as defined in sections 4.3, E_i is the incident electric field, and δ is the phase difference given by:

$$\delta = k\Delta = \frac{2\pi}{\lambda} 2n_e w \cos(\alpha) \tag{5.48}$$

the second beam is reflected twice inside the etalon so that it can be expressed as:

$$\begin{aligned} E_2 &= E_1 \sqrt{\Gamma} e^{-j\delta_r} e^{-jk_0 n_e w} \sqrt{\Gamma} e^{-j\delta_r} \\ &= E_i T e^{j\delta} \Gamma e^{-j(2\delta_r + k_0 n_e w)} \end{aligned} \tag{5.49}$$

where Γ is the reflectivity coefficient as defined in section 4.3.

A third beam, and subsequent beams can be expressed as:

$$\begin{aligned} E_n &= E_{(n-1)} \Gamma e^{-j(2\delta_r + k_0 n_e w)} \\ &= E_0 T e^{j\delta} [\Gamma e^{-j(2\delta_r + k_0 n_e w)}]^n \end{aligned} \tag{5.50}$$

Adding all electric fields we obtain:

$$\frac{E_t}{E_i} = T e^{j\delta} (1 + \Gamma e^{-j(2\delta_r + k_0 n_e w)} + [\Gamma e^{-j(2\delta_r + k_0 n_e w)}]^2 + \cdots) \tag{5.51}$$

we finally need to calculate the intensity which is the observable quantity on the interferometer. From [1, 2], we can calculate the ratio of the transmitted intensity with respect to the incident intensity as a function of the reflectivity of the etalon as:

$$\frac{I_t}{I_i} = \left(\frac{\mathcal{T}}{1 - \mathcal{R}}\right)^2 \cdot \frac{1}{1 + [4\mathcal{R}/(1 - \mathcal{R})^2] \sin^2(\delta/2)} \tag{5.52}$$

where $\mathcal{T}$ and $\mathcal{R}$ are the transmittance and reflectance as defined in section 4.3. In the case of no absorption, we have $\mathcal{R} + T = 1$, and equation (5.52), can be simplified to:

$$\frac{I_t}{I_i} = \mathcal{A} = \frac{1}{1 + [4\mathcal{R}(1 - \mathcal{R})^2] \sin^2(\delta/2)} \tag{5.53}$$

where $\mathcal{A}$ is the Airy function. The Airy function will helps us describe an ideal multiple-beam interferometer with reflectance $\mathcal{R}$. In practice, the Airy profile will be modified by non-ideal mirrors.

We can rewrite the Airy function in terms of a new term, the *Finesse Coefficient*, defined as:

$$F = \frac{4\mathcal{R}}{(1 - \mathcal{R})^2} \tag{5.54}$$

and the Airy function would be then:

$$\mathcal{A} = \frac{1}{1 + F \sin^2(\delta/2)} \tag{5.55}$$

A plot of the Airy function is shown in figure 5.12. The phase separation between adjacent transmittance peaks is sometimes called the free spectral range (FSR) of the etalon. For a plane-parallel plate, the fringes, in transmitted light, will consist of a series of narrow bright rings on an almost completely dark background.

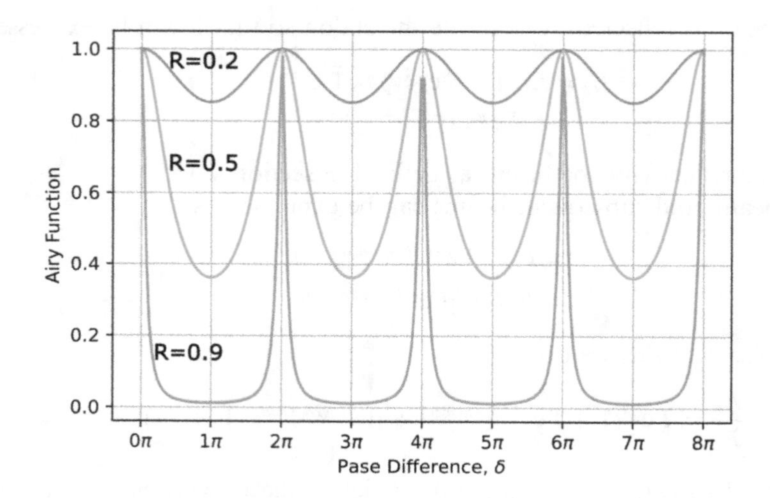

(a) Airy function as a function of phase difference for selected values of $\mathcal{R}$.

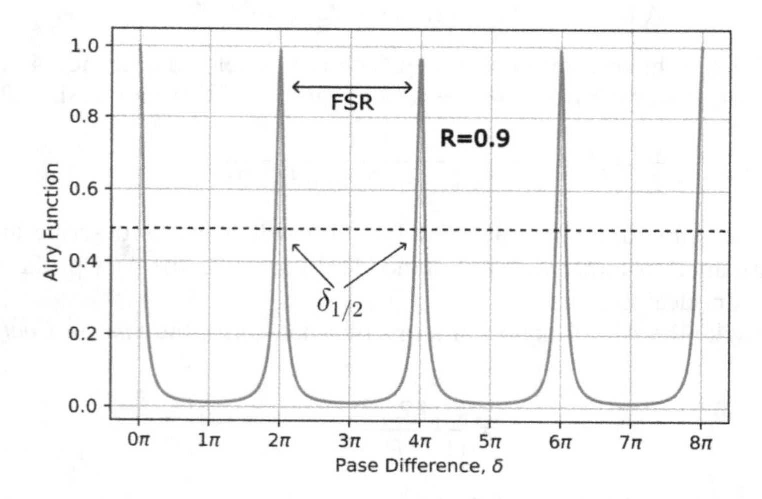

(b) Free Spectral Range and half-width transmittance .

Figure 5.12. Fabry–Perot interferometer fringe profile, as a function of phase difference.

In figure 5.12, we show the phase separation between adjacent transmittance peaks. This is called the *Free Spectral Range* (FSR), and the half-width of the transmittance peaks can be found when $\mathcal{A} = 1/2$, that is:

$$\frac{1}{2} = \frac{1}{1 + F\sin^2 \delta/2} \tag{5.56}$$

solving for δ, we get:

$$\delta_{1/2} = 2\arcsin\left(\frac{1}{\sqrt{F}}\right) \tag{5.57}$$

For large values of F, we can approximate $\arcsin(1/\sqrt{F}) \approx 1/\sqrt{F}$, and simplify $\delta_{1/2}$ to just:

$$\delta_{1/2} = \frac{2}{\sqrt{F}} \tag{5.58}$$

Locating the $\delta_{1/2}$, helps us define the *full-width at half maximum* (FWHM). The FWHM is the distance between the two points on either side of the peak where the signal or intensity is equal to half its maximum value. The FWHM is a measure of the width or broadness of the peak or curve, and it is affected by various factors, such as the resolution of the measuring instrument, the physical properties of the sample, and the underlying physics of the phenomenon being studied. The FWHM is often used as a standard parameter for quantifying the spectral or signal resolution of an instrument or system. In many applications, a narrower FWHM indicates higher resolution or better performance, while a broader FWHM indicates lower resolution or poorer performance. For example, in a spectroscopic measurement, a narrower FWHM indicates higher spectral resolution and greater ability to distinguish between closely spaced spectral features.

If we, for example, use a light source with two wavelengths, λ_1 and λ_2. They will each produce a maximum peak, with their own FWHM. In order to be able to distinguish each wavelength we need to satisfy what is called the *Rayleigh criterion.* The Rayleigh criterion states that the two sources are just barely distinguishable from each other when the maximum of one source's interference pattern overlaps with the minimum of the other source's interference pattern. We can use the value of $\delta_{1/2}$ to find this minimum. Figure 5.13 illustrates this.

The resolving power of a Fabry–Perot interferometer can be calculated as:

$$\left|\frac{\lambda}{\Delta\lambda}\right| = \left[\frac{\pi\sqrt{F}}{2}\right]\frac{2n_e w}{\lambda} \tag{5.59}$$

We can then solve for the finesse coefficient as:

$$F = \left[\frac{1}{\pi n_e w}\frac{\lambda^2}{\Delta\lambda}\right]^2 \tag{5.60}$$

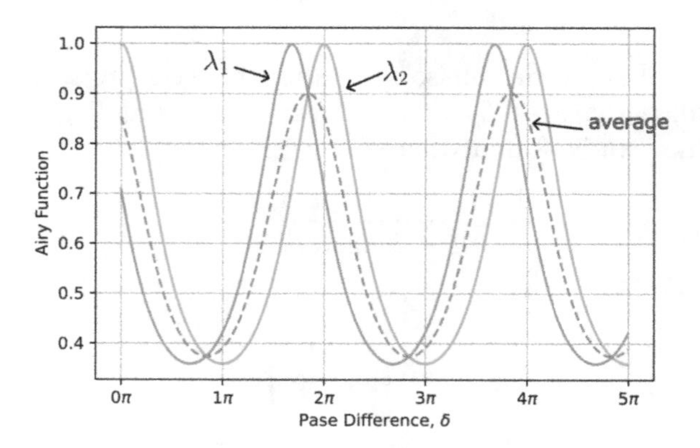

(a) Wavelengths are too close, and won't be able to be resolved.

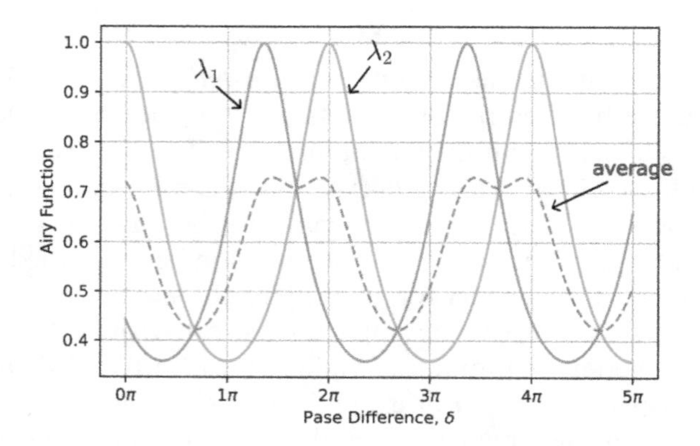

(b) Both wavelengths can be resolved due to the dip on the average intensity

Figure 5.13. Two different wavelengths in a Fabry–Perot interferometer, with $R = 0.7$.

Example 5.3. Calculate the reflectivity needed in a 1 cm etalon, in order to have a resolution of 5×10^{-4} nm. Assume a central wavelength of 600 nm, and air between the etalon's surface.

We first calculate the finesse coefficient.

$$
\begin{aligned}
F &= \left[\frac{1}{\pi n_{\mathrm{e}} w} \frac{\lambda^2}{\Delta\lambda} \right]^2 \\
&= \left[\frac{1}{1 \times 10^{-2} \pi} \frac{(600 \times 10^{-9})^2}{5 \times 10^{-13}} \right]^2 \\
&= 525.24
\end{aligned}
$$

From our definition of finesse coefficient we can then find our value of reflectance.

$$F = \frac{4\mathcal{R}}{(1-\mathcal{R})^2}$$

Using numerical methods, we obtain two possible solutions $\mathcal{R} = 1.0911$ but is not possible to have reflectances higher than 1, and $\mathcal{R} = 0.916\,45$, which will be our only possible solution. Finally, our reflectivity will be the square root of our reflectance, so we get:

$$\Gamma = 0.957$$

An application of Fabry–Perot interferometers is in spectroscopy. They are used to measure the spectral properties of light, such as its intensity and wavelength distribution. The basic principle of using a Fabry–Perot interferometer in spectroscopy is to pass the light through the interferometer and observe the resulting interference pattern. The interferometer is basically used as a narrowband filter to select a specific wavelength or range of wavelengths from a broadband source, such as a lamp or a laser. The interference pattern generated by the interferometer can be used to measure the intensity of the selected wavelength(s) as a function of the mirror spacing. By varying the mirror spacing, a series of interference patterns can be obtained, which can be used to construct a spectrum of the source.

The Fabry–Perot interferometer is also a key component in the design of many types of lasers, including semiconductor lasers, gas lasers, and solid-state lasers. It is used to create a resonant cavity that enhances the intensity and coherence of the laser beam, resulting in a more efficient and stable laser output. They are also used in fiber-optic communication systems to filter out unwanted wavelengths of light and to stabilize the frequency of laser sources. They can also be used to measure the reflectivity and transmission characteristics of optical fibers.

References

[1] Hecht E 2012 *Optics* (London: Pearson)

[2] Pedrotti F L, Pedrotti L M and Pedrotti L S 2017 *Introduction to Optics* (Cambridge: Cambridge University Press)

[3] Robinson A 2006 *The Last Man Who Knew Everything: Thomas Young, the Anonymous Polymath Who Proved Newton Wrong, Explained How We See, Cured the Sick, and Deciphered the Rosetta Stone, Among Other Feats of Genius* (A Plume book) (Pi Press)

IOP Publishing

Optical Sensors
An introduction with lab demonstrations
Victor Argueta-Diaz

Chapter 6

Diffraction

6.1 Introduction

When we talked about double-slit interference, the irradiance equation (5.30) that we got for our interference pattern was a function of the several physical parameters of the setup, but the size of slits didn't appear in this equation. In order to understand the effects of the aperture's size and shape it is necessary to treat the double-slit setup as a diffraction problem. Optical diffraction is a phenomenon that occurs when a wave of light encounters an obstacle or passes through an aperture that is comparable in size to its wavelength. The wave of light diffracts, or bends, around the obstacle or aperture, creating a pattern of bright and dark regions on a screen placed behind it. In this section we will focus on two kinds of diffraction: Fraunhofer diffraction, also known as far-field diffraction, and Fresnel diffraction, or near-field, diffraction.

Near-field diffraction refers to diffraction that occurs in the near-field region, which is the region close to the diffracting object or aperture. In this region, the distance between the diffracting object and the screen is typically smaller than the wavelength of the diffracting wave. Near-field diffraction patterns are typically complex and difficult to analyze because they depend on the detailed structure of the diffracting object, and they exhibit strong spatial variations in intensity and phase.

Far-field diffraction, on the other hand, refers to diffraction that occurs in the far-field region, which is the region far away from the diffracting object or aperture. In this region, the distance between the diffracting object and the screen is much larger than the wavelength of the diffracting wave. Far-field diffraction patterns are simpler and easier to analyze than near-field patterns because they are less sensitive to the detailed structure of the diffracting object. They exhibit regular patterns of bright and dark fringes, and the position and spacing of these fringes can be used to determine various properties of the diffracting object, such as its size, shape, and orientation.

doi:10.1088/978-0-7503-4876-8ch6

6.2 Babinet's principle

During most of this chapter, we will discuss 'apertures' as diffractive elements. However, we can use Babinet's principle to obtain the diffraction pattern caused by obstructions.

Babinet's principle is a physics principle that states that the diffraction pattern produced by a perfectly conducting, opaque body is identical to that produced by a hole of the same size and shape in an infinitely thin, perfectly conducting screen. In other words, the diffraction patterns produced by an object and its complementary object (which has the same shape but is empty where the object is solid and vice versa) are the same.

For example, if you have a solid rectangle and shine a beam of light through it, the resulting diffraction pattern will be the same as if you had a rectangular hole of the same size and shape and shone a beam of light through it. This principle can be applied to any object and its complementary object, regardless of their shape, as shown in figure 6.1.

6.3 Huygens–Fresnel principle

We'll start our discussion by explaining the Huygens–Fresnel principle. The Huygens–Fresnel principle is a fundamental concept in wave optics that describes how light waves propagate through space. It states that every point on a given wavefront can be considered as a source of secondary wavelets that propagate in all directions. These secondary wavelets, in turn, become new sources of waves, producing a new wavefront that is tangent to the secondary wavelets.

This principle was first proposed by Dutch physicist Christiaan Huygens in the late 17th Century as a way to explain the propagation of light as a wave phenomenon. He argued that the wavefront of light can be considered as a surface made up of an infinite number of point sources, each of which generates a new spherical wave. The sum of all these waves creates a new wavefront that propagates through space.

Later, French physicist Augustin-Jean Fresnel expanded on Huygens' ideas by adding the concept of wave interference. According to Fresnel, the actual field at any

Figure 6.1. The diffraction pattern of an aperture is the same as the diffraction pattern of an obstruction with the same shape.

point beyond the wavefront is a superposition of all the secondary wavelets, taking into account both their amplitudes and phases. This means that the interference of waves plays a critical role in determining the behavior of light waves.

Consider an arbitrary aperture located at the plane $z = 0$. Let the aperture be illuminated with a field distribution $E(x', y', z')$ in the aperture. Then for a point $P(x, y, z)$, away from the aperture, the net field is given by adding together the contribution of wavelets emitted from each point in the aperture. This can be expressed as follows [1]:

$$E_p(x, y, z) = \frac{1}{j\lambda} \iint_{\text{area}} \left[\frac{1 + \cos(\theta)}{2}\right] \frac{\exp(jkr)}{r} E(x', y', z') dx'dy' \tag{6.1}$$

where $r = \sqrt{(x - x')^2 + (y - y')^2 + z^2}$, and the term in brackets is known as the *obliquity factor*. $\cos(\theta) = z/r$, and θ is the angle between the normal to the aperture (z-axis in our case), and the unit vector $\hat{r}$ that's parallel to the trajectory r.

We can also rewrite r for the numerator as:

$$\begin{aligned} r &= z\sqrt{1 + \frac{(x - x')^2 + (y - y')^2}{z^2}} \\ &\approx z\left[1 + \frac{(x - x')^2 + (y - y')^2}{2z^2} - \frac{1}{8}\left(\frac{(x - x')^2 + (y - y')^2}{z^2}\right)^2 + \cdots\right] \end{aligned} \tag{6.2}$$

and for the denominator we will assume $r = z$. We cannot use the same approximation for r in the numerator because it is multiplied by a large value of k, which is equal to $2\pi/\lambda$. Any error introduced by approximating r would be significantly magnified by k.

Taking only the first two terms of r and substituting it in equation (6.1), we get:

$$\begin{aligned} E_p(x, y, z) = &\frac{1}{jz\lambda}\left[\exp(jkz)\exp\left(\frac{jk}{2z}(x^2 + y^2)\right)\right] \\ &\times \iint_{\text{area}} \left[\frac{1 + \cos(\theta)}{2}\right] \exp\left(\frac{jk}{2z}(x'^2 + y'^2)\right) \exp\left(\frac{-jk}{2z}(xx' + yy')\right) \\ &\times E(x', y', z')dx'dy' \end{aligned} \tag{6.3}$$

6.4 Fraunhofer diffraction

When working in the far-field region, we will assume that $z > a^2/\lambda$. where a is the largest dimension of the aperture. Under this condition, the obliquity factor is close to one, and one of the exponential factors is also close to one.

$$\exp\left(\frac{jk}{2z}(x'^2 + y'^2)\right) \approx 1 \tag{6.4}$$

We can rewrite equation (6.1), as:

$$E_p(x, y, z) = \frac{1}{jz\lambda}\left[\exp(jkz)\exp\left(\frac{jk}{2z}(x^2 + y^2)\right)\right] \times \int\int_{\text{area}} \exp\left(\frac{-jk}{2z}(xx' + yy')\right)E(x', y', z')dx'dy' \tag{6.5}$$

Which is called the Fraunhofer approximation for diffraction. Let's look at the diffraction pattern created by a single rectangular aperture, as shown in figure 6.2. We will assume that the length of the aperture, b, and its width, a, are of similar dimensions and that the aperture is being illuminated by a plane wave.

Using the Fraunhofer approximation to calculate the electric field at point P, we obtain:

$$E_p(x, y, z) = C\int_{-a/2}^{a/2} \exp\left(\frac{-jk}{2z}xx'\right)dx' \int_{-b/2}^{b/2} \exp\left(\frac{-jk}{2z}yy'\right)dy' \tag{6.6}$$

where:

$$C = \frac{E_0}{jz\lambda}\left[\exp(jkz)\exp\left(\frac{jk}{2z}(x^2 + y^2)\right)\right] \tag{6.7}$$

is used to simplify the expression. Integrating the Fraunhofer diffraction, we obtain:

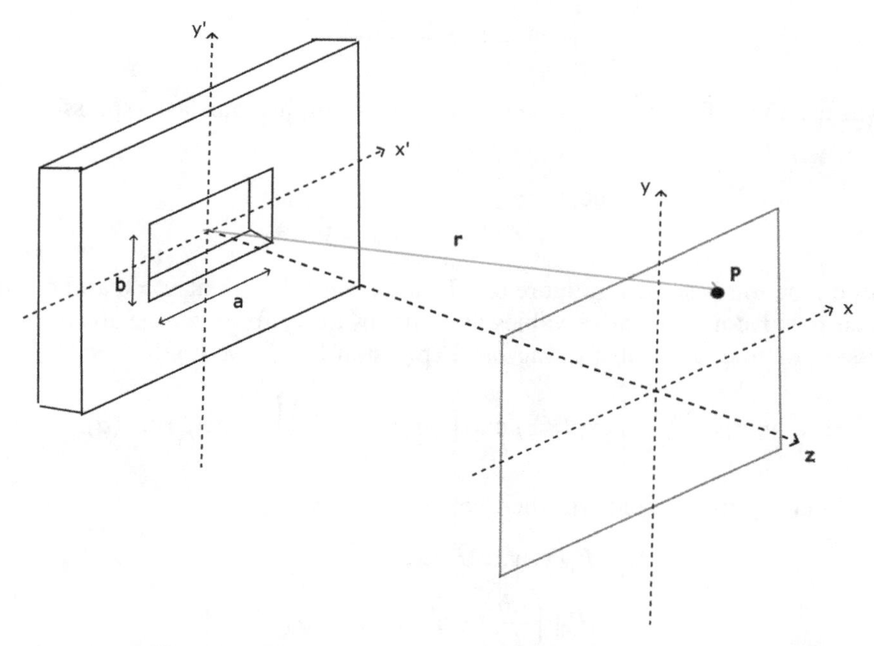

Figure 6.2. Rectangular aperture illuminated by a plane wave.

$$E_p(x, y, z) = C\frac{a}{2}\frac{\exp\left(\frac{jkax}{2z}\right) - \exp\left(\frac{-jkax}{2z}\right)}{jkx} \times \frac{b}{2}\frac{\exp\left(\frac{jkby}{2z}\right) - \exp\left(\frac{-jkby}{2z}\right)}{jky} \tag{6.8}$$

by defining:

$$\beta = \frac{kax}{2z} \tag{6.9}$$

and,

$$\alpha = \frac{kby}{2z} \tag{6.10}$$

we can rewrite (6.8) as:

$$E_p(x, y, z) = Ca\frac{\exp(j\beta) - \exp(-j\beta)}{j2\beta} \times b\frac{\exp(j\alpha) - \exp(-j\alpha)}{j2\alpha} \tag{6.11}$$

by using Euler's identities we can further simplified to:

$$E_p(x, y, z) = Cab\frac{\sin(\beta)}{\beta}\frac{\sin(\alpha)}{\alpha} \tag{6.12}$$

We will use the definition of the sinc function to simplify our last expression.

$$\mathrm{sinc}(x) = \begin{cases} \frac{\sin(x)}{x}, & x \neq 0 \\ 1, & x = 0 \end{cases} \tag{6.13}$$

The sinc function is shown in figure 6.3. It has a central peak at $x = 0$ and oscillates between positive and negative values as x moves away from zero. Substituting the expression of sinc, and substituting our expression for C, we finally get:

$$E_p(x, y, z) = \frac{abE_0}{jz\lambda}\left[\exp(jkz)\exp\left(\frac{jk}{2z}(x^2 + y^2)\right)\right] \cdot \mathrm{sinc}(\beta)\mathrm{sinc}(\alpha) \tag{6.14}$$

The intensity of the field will then be:

$$\begin{aligned} I_p &= E_p(x, y, z)\overline{E_p(x, y, z)} \\ &= |E_0|^2\left(\frac{ab}{\lambda z}\right)^2 \mathrm{sinc}^2(\beta)\mathrm{sinc}^2(\alpha) \end{aligned} \tag{6.15}$$

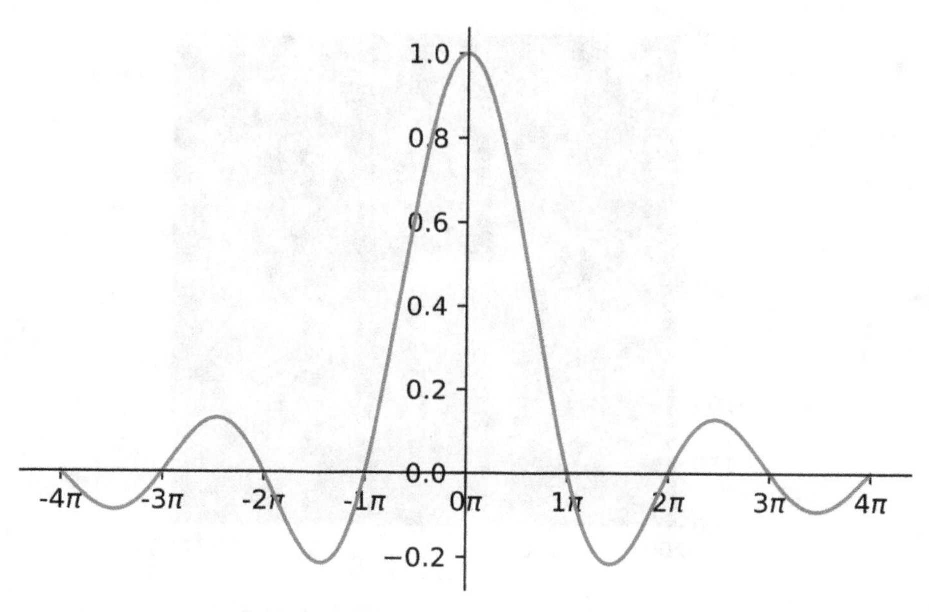

Figure 6.3. Sinc function as a function of β.

where $\overline{E_p(x, y, z)}$ indicates the complex conjugate of the function. We can see from the figure 6.3 that the sinc function crosses the x-axis at $m\pi$, except for $m = 0$. These points will represent dark fringes in our diffraction pattern. We can find these points for the x-axis from our definition of β:

$$\begin{aligned} \beta &= \frac{kax}{2z} = m\pi \\ &= \frac{1}{2}\frac{2\pi}{\lambda}\frac{ax}{z} = m\pi \\ \frac{ax}{z} &= m\lambda \end{aligned} \tag{6.16}$$

and finally we can get an expression for the location of minimum values (dark fringes) of our diffraction pattern

$$x_m = \frac{m\lambda z}{a} \tag{6.17}$$

similarly, for the y-direction we get:

$$y_n = \frac{n\lambda z}{b} \tag{6.18}$$

where m, and n are integers values different than zero.

Figure 6.4, shows the intensity of a diffraction pattern for a rectangular aperture with dimensions 20 μm × 10 μm for a wavelength of 500 nm.

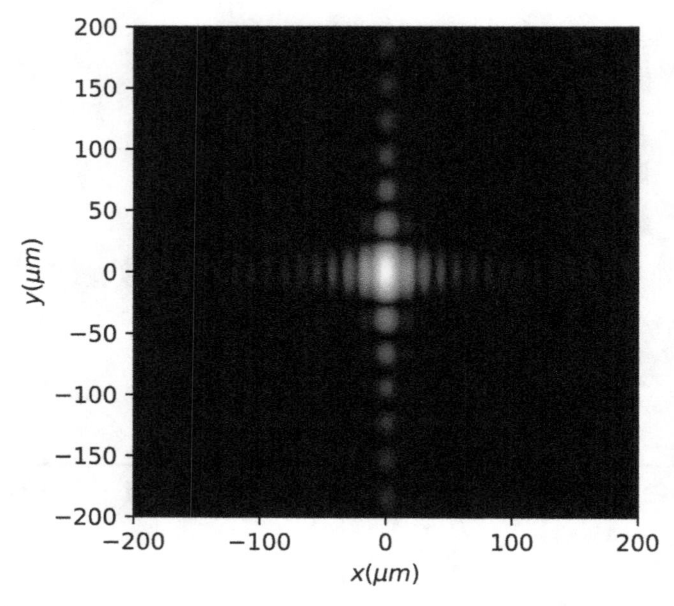

Figure 6.4. Simulation of the diffraction pattern for a rectangular aperture of dimensions $a = 20\ \mu\text{m}$ and $b = 10\ \mu\text{m}$.

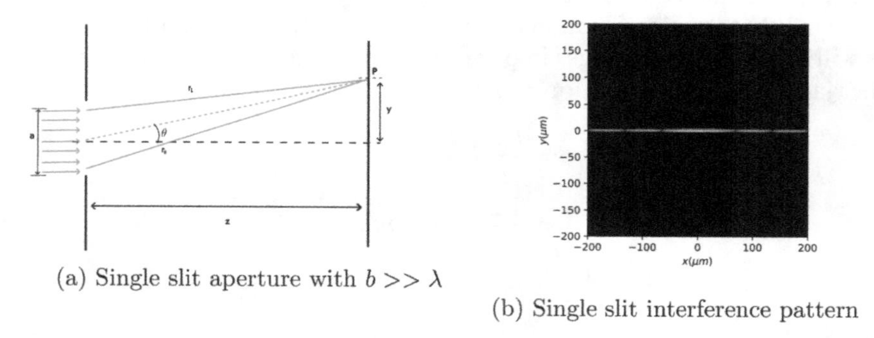

(a) Single slit aperture with $b >> \lambda$

(b) Single slit interference pattern

Figure 6.5. Diffraction pattern for a slit with aperture $a = 5\ \mu\text{m}$ and a wavelength of 500 nm.

Let's look now at the diffraction pattern created by a single slit such as the one shown in figure 6.5. We assume that $b \gg \lambda$ so that the diffraction effects on this dimension can be neglected. We will use our previous results, at $y = 0$, to obtain the diffraction interference pattern for this aperture. We can write our electric field as:

$$E_{\mathrm{p}}(x, y, z) = C\int_{-a/2}^{a/2} \exp\left(\frac{-jk}{2z}xx'\right)dx' \tag{6.19}$$

we are using again equation (6.7) to simplify our expression. Solving for the integral we obtain:

$$E_p(x, y, z) = Ca\frac{\exp\left(j\frac{kax}{2z}\right) - \exp\left(-j\frac{kax}{2z}\right)}{j2\frac{kax}{2z}} = Ca \cdot \text{sinc}\left(\frac{kax}{2z}\right) \tag{6.20}$$

The intensity for a single-slit aperture will then be:

$$I_p = |E_0|^2\left(\frac{a}{\lambda z}\right)^2 \text{sinc}^2\left(\frac{kax}{2z}\right) \tag{6.21}$$

The size of the central bright fringe will be the distance between the x_{-1} and x_1 dark fringes, we can calculate this distance using equation (6.17), that is:

$$\Delta x = x_1 - x_{-1} = \frac{2\lambda z}{a} \tag{6.22}$$

We can see from our last equation that the smaller our slit is, the larger the main bright fringe will be.

6.4.1 Circular aperture

The calculation of the diffraction pattern created by a circular aperture follows a similar procedure to that done for the rectangular slit. However, we need to modify equation (6.5), to cylindrical coordinates and integrate for the area of the circular aperture. We will use the following identities to convert our equation to cylindrical coordinates:

$$\rho = \sqrt{x^2 + y^2} \tag{6.23}$$

$$x = \rho \cos \theta \tag{6.24}$$

$$y = \rho \sin \theta \tag{6.25}$$

and similar expressions for the prime coordinates. Using these identities, our Fraunhofer equation can be written as:

$$\begin{aligned} E_p(\rho, z) = &\frac{1}{jz\lambda}\left[\exp(jkz)\exp\left(\frac{jk\rho^2}{2z}\right)\right] \\ &\times \int\int \exp\left(\frac{jk\rho'^2}{2z}\right) \\ &\times \exp\left(\frac{jk}{z}(\rho\rho' \cos(\phi - \phi'))\right)E(\rho', 0)d\phi'\rho' d\rho' \end{aligned} \tag{6.26}$$

We'll take a close look at ϕ-dependent exponential. If we assume a symmetric system around ϕ we can solve the integral by assuming $\phi = 0$. The solution of the resultant integral is called the Bessel function of the first order. Bessel functions are a family of mathematical functions, such as trigonometric functions, that arise in a wide range of physical and engineering applications, particularly in problems that involve wave propagation, such as heat transfer, acoustics, electromagnetism, and quantum mechanics. We will explain further about Bessel functions in section 7.4. For now, we write a definition to calculate the Bessel function as:

$$\int_0^{2\pi} \exp\left(\frac{jk}{z}(\rho\rho' \cos(\phi))\right)d\phi' = 2\pi J_0\left(\frac{k\rho\rho'}{z}\right) \tag{6.27}$$

and using our Bessel function into equation (6.26), we get:

$$\begin{aligned} E_{\mathrm{p}}(\rho, z) = \frac{2\pi}{jz\lambda}\left[\exp(jkz)\exp\left(\frac{jk\rho^2}{2z}\right)\right] \\ \times \int \exp\left(\frac{jk\rho'^2}{2z}\right)J_0\left(\frac{k\rho\rho'}{z}\right)E(\rho', 0)\rho' d\rho' \end{aligned} \tag{6.28}$$

For the Fraunhofer approximation:

$$\exp\left(\frac{jk\rho'^2}{2z}\right) \approx 1 \tag{6.29}$$

and finally, we get an expression for the Fraunhofer diffraction in cylindrical coordinates.

$$E_{\mathrm{p}}(\rho, z) = \frac{2\pi}{jz\lambda}\left[\exp(jkz)\exp\left(\frac{jk\rho^2}{2z}\right)\right]\int J_0\left(\frac{k\rho\rho'}{z}\right)E(\rho', 0)\rho' d\rho' \tag{6.30}$$

The solution of this equation will depend on the specifics of our electric field.

Example 6.1. Calculate the diffraction pattern for a circular aperture of radius a, illuminated by a plane wave.

$$E_{\mathrm{p}}(\rho, z) = \frac{2\pi E_0}{jz\lambda}\left[\exp(jkz)\exp\left(\frac{jk\rho^2}{2z}\right)\right]\int_0^a J_0\left(\frac{k\rho\rho'}{z}\right)\rho' d\rho' \tag{6.31}$$

to solve this integral we use the integral identity for Bessel functions of the first order:

$$\int_0^u u' J_0(u')du' = uJ_1(u) \tag{6.32}$$

let's set $u = \frac{k\rho\rho'}{z}$, we get

$$\begin{aligned} \int_0^a J_0\left(\frac{k\rho\rho'}{z}\right)\rho' d\rho' &= \left(\frac{z}{k\rho}\right)^2 \int_{u=0}^{u=ka\rho/z} uJ_0(u)du \\ &= \frac{za}{k\rho}J_1\left(\frac{k\rho a}{z}\right) \end{aligned} \tag{6.33}$$

substituting equation (6.33) into equation (6.31), we get:

$$E_p(\rho, z) = \frac{2\pi a E_0}{jk\rho\lambda}\left[\exp(jkz)\exp\left(\frac{jk\rho^2}{2z}\right)\right]J_1\left(\frac{k\rho a}{z}\right) \quad (6.34)$$

and calculating the field intensity ($E_p\overline{E_p}$), we get:

$$I_p = |E_0|^2\left(\frac{\pi a}{2\lambda z}\right)^2\left[2\frac{J_1(ka\rho/z)}{ka\rho/z}\right]^2 \quad (6.35)$$

The term in square brackets is called the *jinc* function, it behaves similarly to our sinc function, but the first zero for $J_1(u) = 0$, will be at $u = 3.83$.

It is possible to find the radius of this dark fringe from $\frac{k\rho a}{z} = 3.88$, and $k = 2\pi/\lambda$, as:

$$\rho_1 = 1.22\frac{z\lambda}{2a} \quad (6.36)$$

This bright central bright spot surrounded by a series of concentric rings is called the *Airy disk*. The diameter of the central spot is determined by the size of the aperture and the wavelength of the light. The rings around the Airy disc are called diffraction rings, and their intensity decreases with increasing distance from the center.

The Airy disk has important implications for imaging and microscopy because it determines the minimum resolvable distance between two closely spaced objects. When two objects are closer together than the diameter of the Airy disk, they cannot be resolved as separate entities and appear as a single blurred image.

Figure 6.6 shows the diffraction pattern for a circular aperture with a 25 μm radius, illuminated by a plane wave with a wavelength of 500 nm, and observed at a distance $z = 10$ mm. We get an Airy disk with a radius of 122 μm.

6.4.2 Multiple slits diffraction

We so far have dealt with individual apertures. It is possible to calculate the diffraction caused by multiple apertures. These structures are called *diffraction gratings*, and usually have a very large number of diffraction elements. To better understand how to calculate the diffraction pattern caused by a large number of elements, it may be beneficial to look at the case of just two apertures.

Figure 6.7, shows a plane with two slits of equal dimension, separated by a distance *d*. This is pretty much the same configuration that we used when looking at Young's experiment in section 5.2.1.

We will use the Fraunhofer approximation shown in equation (6.5) again, and integrate along the *y*-axis for our two apertures. We will split the *y*-direction integral into two parts, one part will integrate for the first aperture, the second part will integrate over the second aperture:

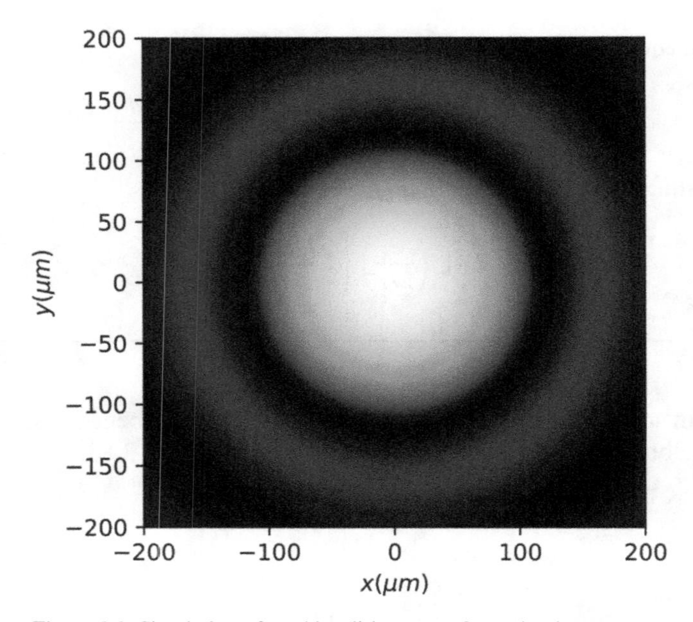

Figure 6.6. Simulation of an Airy disk pattern for a circular aperture.

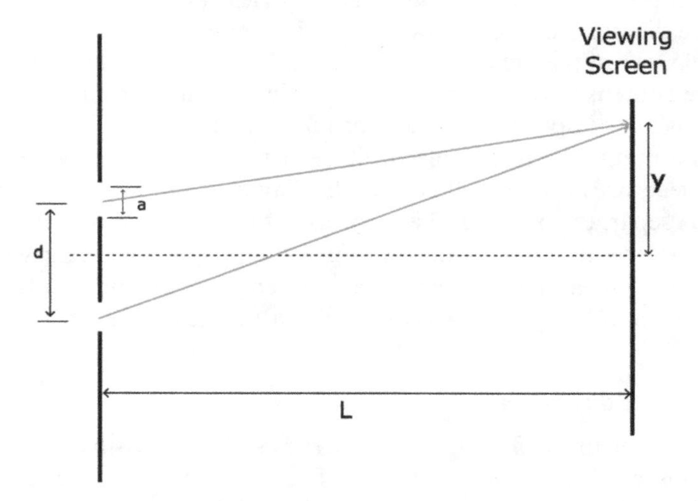

Figure 6.7. Double-slit configuration. Both slits are separated by a distance d, and have the same aperture a.

$$E_{\mathrm{p}}(x, y, z) = \frac{E_0}{jz\lambda}\left[\exp(jkz)\exp\left(\frac{jk}{2z}(x^2 + y^2)\right)\right] \times\left[\int_{-(1/2)(d+a)}^{-(1/2)(d-a)} \exp\left(\frac{-jk}{2z}yy'\right)dy' + \int_{(1/2)(d-a)}^{(1/2)(d+a)} \exp\left(\frac{-jk}{2z}yy'\right)dy'\right] \tag{6.37}$$

Solving for our first integral we get:

$$\int_{-(1/2)(d+a)}^{-(1/2)(d-a)} \exp\left(\frac{-jk}{2z}yy'\right)dy' = \frac{\exp\left(\frac{jk(a-d)x}{2z}\right) - \exp\left(\frac{-jk(a+d)x}{2z}\right)}{jkx} \tag{6.38}$$

and for our second integral, we get:

$$\int_{(1/2)(d-a)}^{(1/2)(d+a)} \exp\left(\frac{-jk}{2z}yy'\right)dy' = \frac{\exp\left(\frac{jk(a+d)x}{2z}\right) - \exp\left(\frac{-jk(d-a)x}{2z}\right)}{jkx} \tag{6.39}$$

we can do a variable substitution in terms of d, and a, as follows:

$$\beta = \frac{kax}{2z} \tag{6.40}$$

and,

$$\alpha = \frac{kdx}{2z} \tag{6.41}$$

and we can rewrite equations (6.37) as,

$$\begin{aligned} E_{\mathrm{p}}(x, y, z) = \frac{E_0}{jz\lambda}&\left[\exp(jkz)\exp\left(\frac{jk}{2z}(x^2+y^2)\right)\right] \\ &\times \frac{a}{2j\beta}[\mathrm{e}^{j\alpha}(\mathrm{e}^{j\beta}-\mathrm{e}^{-j\beta}) + \mathrm{e}^{-j\alpha}(\mathrm{e}^{j\beta}-\mathrm{e}^{-j\beta})] \end{aligned} \tag{6.42}$$

which can be simplified using Euler's identities to:

$$\begin{aligned} E_{\mathrm{p}}(x, y, z) = \frac{E_0}{jz\lambda}&\left[\exp(jkz)\exp\left(\frac{jk}{2z}(x^2+y^2)\right)\right] \\ &\times \frac{a}{2j\beta}(2j\sin(\beta))(2\cos(\alpha)) \end{aligned}$$

We finally get,

$$\begin{aligned} E_{\mathrm{p}}(x, y, z) = \frac{E_0}{jz\lambda}&\left[\exp(jkz)\exp\left(\frac{jk}{2z}(x^2+y^2)\right)\right] \\ &\times (2a)\mathrm{sinc}(\beta)\cos(\alpha) \end{aligned} \tag{6.43}$$

We can finally write the intensity of the diffraction pattern as a function of the separation of the slits and their aperture dimension. The diffraction pattern is shown in figure 6.8(a), and its irradiance in equation (6.44)

$$I_{\mathrm{p}}(x, y, z) = \left(\frac{2aE_0}{z\lambda}\right)^2 \mathrm{sinc}^2(\beta)\cos^2(\alpha) \tag{6.44}$$

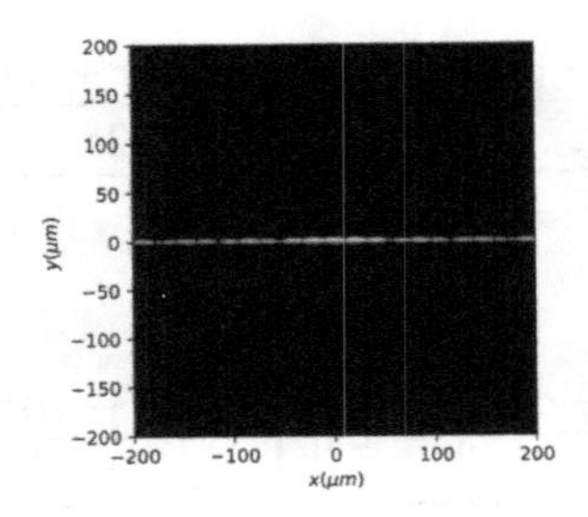

(a) Simulation of the diffraction pattern caused by a double slit

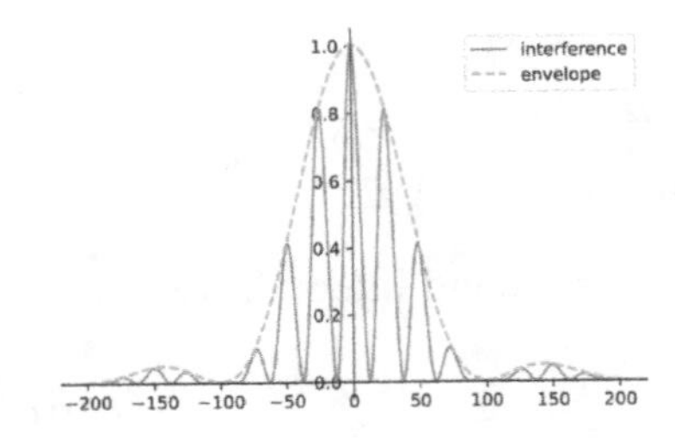

(b) Intensity distribution for the diffraction pattern caused by a double slit

Figure 6.8. Diffraction pattern and Intensity distribution for a double-slit with a slit width of 5 μm, and a spacing of 20 μm.

But what if we have a large number of slits, or the apertures are triangular or some other arbitrary shape? Let's take a closer look at the expressions for the double-slit interference (equation (5.30)), the single-slit diffraction (equation (6.21)), and the double-slit diffraction that we just obtained (equation (6.44)), for the time being, we will ignore the amplitude term:

$$I_{\mathrm{p}} = I_0 \cos^2\left(\frac{kxd}{2z}\right), \qquad \text{double-slit interference}$$

$$I_{\mathrm{p}} = I_0 \mathrm{sinc}^2\left(\frac{kax}{2z}\right), \qquad \text{single-slit diff.}$$

$$I_{\mathrm{p}} = I_0 \mathrm{sinc}^2\left(\frac{kax}{2z}\right)\cos^2\left(\frac{kdx}{2z}\right), \qquad \text{double-slit diff.}$$

We can observe that the third expression is the result of the product of the first two (again, ignoring the amplitude). When we have an array of N-slits of equal shape, we can calculate the diffraction pattern by evaluating the diffraction pattern of a single aperture and multiplying it by the interference pattern of the N-slits. We can see this effect in figure 6.8(b), where we have an interference pattern enveloped in a sinc function. We can express this mathematically as [1]:

$$\begin{aligned} E_{\mathrm{p}}(x, y, z) = &\left[\sum_{n=1}^{N} \exp\left(-j\frac{k}{z}(xx' + yy')\right)\right] \\ &\times\left[\frac{1}{jz\lambda}\exp(jkz)\exp\left(\frac{jk}{2z}(x^2 + y^2)\right)\right. \\ &\left.\iint_{\text{area}} \exp\left(\frac{-jk}{2z}(xx' + yy')\right)E(x', y', z')dx'dy'\right] \end{aligned} \tag{6.45}$$

The second term in brackets is the Fraunhofer diffraction expression of equation (6.5), and the summation on the first bracket represents the contribution caused by N-equal apertures.

Example 6.2. Calculate the diffraction pattern caused by an $N \times M$ array of rectangular apertures of dimension $a \times b$. The array is represented in figure 6.9. The pitch between each aperture is h horizontally, and v vertically. Each aperture will have a location given by:

$$x_n' = \left(n - \frac{N+1}{2}\right)h$$

in the x-direction, and

$$y_m' = \left(m - \frac{M+1}{2}\right)v$$

in the y-direction. We previously calculated the diffraction pattern for a single rectangular aperture, the result is shown in equation (6.15). We now need to calculate the summation component of equation (6.45).

$$\sum_{n,m=1}^{N,M} \exp\left(-j\frac{k}{z}(xx' + yy')\right) = e^{j\frac{khx}{z}\left(\frac{N+1}{2}\right)} \sum_{n=1}^{N} e^{j\frac{khx}{z}n} \times e^{j\frac{kvy}{z}\left(\frac{M+1}{2}\right)} \sum_{m=1}^{M} e^{j\frac{kvy}{z}m}$$

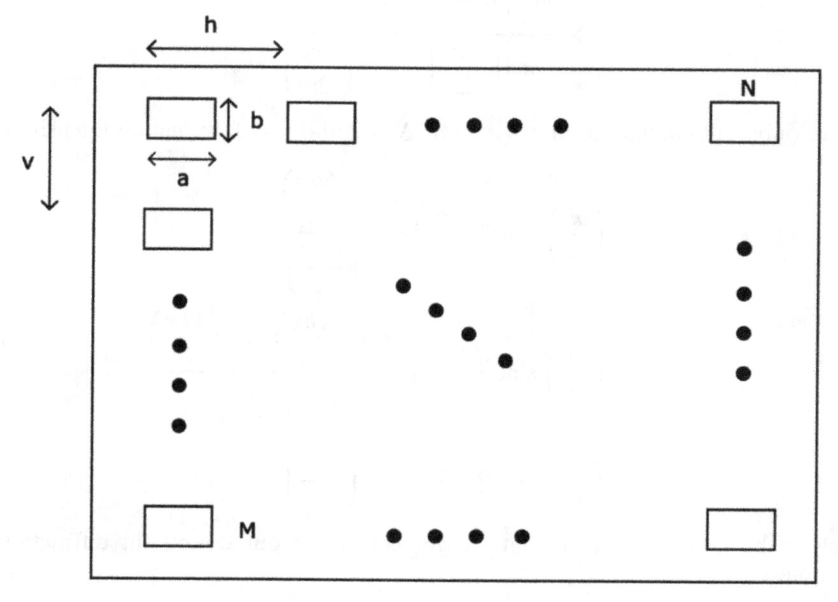

Figure 6.9. 2D array of rectangular apertures.

Looking only at the x-dependent part of the equation we can express the summation as:

$$\sum_{n=1}^{N} \exp\left(-j\frac{k}{z}xx'\right) = e^{j\frac{khx}{z}\left(\frac{N+1}{2}\right)} \frac{e^{-jkhxN/z} - 1}{e^{-jkhx/z} - 1}$$

$$= \frac{e^{-jkhxN/2z} - e^{jkhxN/z}}{e^{-jkhx/2z} - e^{jkhx/2z}}$$

$$= \frac{\sin\left(N\frac{khx}{2z}\right)}{\sin\left(\frac{khx}{2z}\right)}$$

an similarly for the y-dependent summation, we get:

$$\sum_{m=1}^{M} \exp\left(-j\frac{k}{z}yy'\right) = \frac{e^{-jkvyM/2z} - e^{jkvyM/z}}{e^{-jkvy/2z} - e^{jkvy/2z}}$$

$$= \frac{\sin\left(M\frac{kvy}{2z}\right)}{\sin\left(\frac{kvy}{2z}\right)}$$

Finally, we can combine our previous results to get an expression for the intensity:

$$I_\mathrm{p} = |E_0|^2\left(\frac{ab}{\lambda z}\right)^2 \mathrm{sinc}^2\left(\frac{kax}{2z}\right)\mathrm{sinc}^2\left(\frac{kby}{2z}\right) \times \frac{\sin^2\left(N\frac{khx}{2z}\right)}{\sin^2\left(\frac{khx}{2z}\right)} \frac{\sin^2\left(M\frac{kvy}{2z}\right)}{\sin^2\left(\frac{kvy}{2z}\right)}$$

For a double-slit configuration, we assume $N = 2$ and $y = 0$, we get an intensity of:

$$I_\mathrm{p} = |E_0|^2\left(\frac{ab}{\lambda z}\right)^2 \mathrm{sinc}^2\left(\frac{kax}{2z}\right) \frac{\sin^2\left(2\frac{khx}{2z}\right)}{\sin^2\left(\frac{khx}{2z}\right)}$$

$$= |E_0|^2\left(\frac{ab}{\lambda z}\right)^2 \mathrm{sinc}^2\left(\frac{kax}{2z}\right) \frac{2\sin^2\left(\frac{khx}{2z}\right)\cos^2\left(\frac{khx}{2z}\right)}{\sin^2\left(\frac{khx}{2z}\right)}$$

$$= |E_0|^2\left(\frac{ab}{\lambda z}\right)^2 \mathrm{sinc}^2\left(\frac{kax}{2z}\right)\cos^2\left(\frac{khx}{2z}\right)$$

which is the same expression that we got before for our double-slit diffraction in equation (6.44).

6.4.3 Diffraction gratings

When the number of diffractive elements, N, is very high, we say we are working with a diffraction grating. A diffraction grating is an optical device that is used to separate light into its component wavelengths. It consists of a large number of equally spaced parallel slits or lines. When light passes through a diffraction grating, it is diffracted or bent by the slits, causing interference between the waves that are diffracted from each slit.

We will assume that our diffraction grating is formed by an N-number of vertical slits. As we saw in the previous section, the intensity of our interference pattern will be given by:

$$I_{\mathrm{p}} = |E_0|^2\left(\frac{ab}{\lambda z}\right)^2 \mathrm{sinc}^2\left(\frac{kax}{2z}\right) \times \frac{\sin^2\left(N\frac{khx}{2z}\right)}{\sin^2\left(\frac{khx}{2z}\right)}$$

The number of slits has an interesting effect on the diffraction pattern. As the number of slits increases, the diffraction pattern becomes more defined and intense, and the angular separation between the diffracted beams decreases.

The reason for this effect is that, as the number of slits increases, they produce a greater number of interfering wavefronts. The condition for constructive interference needs to be satisfied for all wavefronts, and this only happens in more defined positions. This creates a more pronounced interference pattern with high contrast between the bright and dark fringes of the diffraction pattern.

Figure 6.10 shows the effect of the number of slits on the intensity. We can see that the peaks remain at the same position, but as the number of slits increases, they become sharper. Their intensity also increases to the order of N^2, although we don't see this in our figure because the values are normalized.

The location of the peaks will happen when

$$\frac{khx}{2z} = m\pi \tag{6.46}$$

substituting $k = 2\pi/\lambda$, we find a position of our peaks of:

$$\frac{hx_m}{z} = m\lambda \tag{6.47}$$

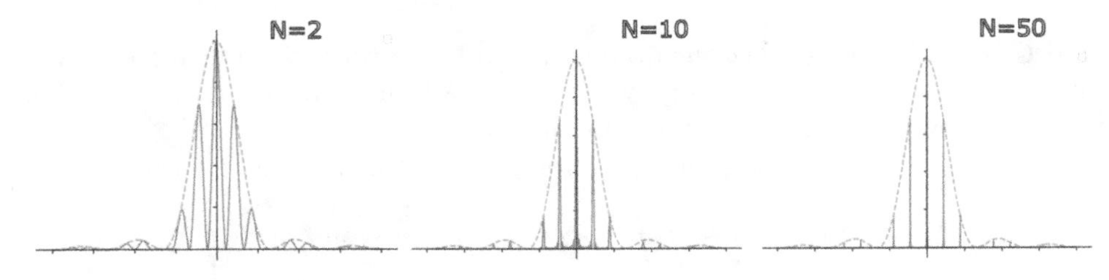

Figure 6.10. Effect of the number of slits on the diffraction pattern.

again, h is the distance between slits, x is the position to our peak, z is the distance from our slits to our viewing screen. This equation is usually written, as

$$d\sin(\theta_m) = m\lambda \tag{6.48}$$

where $\sin(\theta) = x/z$, and we substitute h by the more common variable d.

6.5 Fresnel diffraction

When either the source or the observation screen is close to the aperture, the wavefront curvature cannot be ignored, and the Fraunhofer diffraction approximation may no longer be valid. In these cases, the wavefront curvature must be taken into account, and the full wave theory of diffraction should be used instead of the Fraunhofer approximation. The full wave theory of diffraction takes into account the wavefront curvature and the phase differences between the waves that pass through different parts of the aperture. This theory can be more accurate than the Fraunhofer approximation, but it is also mathematically more complex, and it is usually solved numerically. We already introduced the Fresnel diffraction formula in equation (6.1), which we rewrite here and replaced the obliquity factor with a value of one, which is a fair assumption for forward propagating waves:

$$\begin{aligned} E_{\mathrm{p}}(x, y, z) = &\frac{1}{jz\lambda}\exp(jkz) \\ &\times \iint_{\text{area}} \exp\left(\frac{jk}{2z}((x-x')^2 + (y-y')^2)\right) E(x', y', z')dx'dy' \end{aligned} \tag{6.49}$$

Fresnel diffraction, even with the simplifications that we have done so far, is complicated to solve. There are different techniques on how to solve this integral. Fresnel proposed to separate the aperture in regions called *Fresnel Zones*, which can help solve this integral in simpler equations. For more complex apertures, we can use convolution techniques.

The convolution of two functions f and g is defined as:

$$f(u) * g(u) = \int_0^u f(\tau)g(u-\tau)d\tau \tag{6.50}$$

Comparing our definition of convolution with equation (6.49), we can express Fresnel diffraction as:

$$E_{\mathrm{p}}(x, y, z) = C \cdot E(x, y, z) * \exp\left(\frac{jk}{2z}(x^2 + y^2)\right) \tag{6.51}$$

where C is the factor outside our double integral. In order to avoid working with convolutions, we can, alternatively, work with the Fourier transformations, which will convert the convolution into a multiplication.

The Fourier transformation of a function $g(x, y)$ is defined as:

$$\mathcal{G}(p, q) = \mathcal{F}(g(x, y)) = \iint_{-\infty}^{\infty} g(x, y)\exp(j2\pi(px + qy))dxdy \tag{6.52}$$

where p, and q are spatial frequencies defined as $p = \frac{x}{\lambda z}$, and $q = \frac{y}{\lambda z}$.

To solve the convolution of equation (6.51), we just need to multiply the Fourier transformation of each function:

$$E_p(p, q) = C \cdot \mathcal{F}(E(x, y))\mathcal{F}\left(\exp \frac{jk}{2z}(x^2 + y^2)\right) \tag{6.53}$$

The Fourier transformation of the second term is just

$$\mathcal{F}\left(\exp \frac{jk}{2z}(x^2 + y^2)\right) = \exp \frac{jz}{2k}(p^2 + q^2) \tag{6.54}$$

So the Fresnel diffraction using Fourier transformation is:

$$E_p(p, q) = C \cdot \mathcal{F}(E(x, y))\left(\exp \frac{jz}{2k}(p^2 + q^2)\right) \tag{6.55}$$

Unfortunately, the topic of Fourier optics is beyond the scope of this book. Readers interested in this topic should consult John Goodman's *Introduction to Fourier Optics* [2].

An interesting diffraction effect is what is called the *Poisson point*. The Poisson point, also known as the Arago spot, is a bright spot that appears at the center of a circular shadow cast by a small opaque object illuminated by a distant point source of light, such as the Sun or a star.

According to classical wave theory of light, the shadow of a circular object should be completely dark in the center, as the waves of light are blocked by the object. However, François Arago, a French astronomer and physicist first observed in 1810 that a bright spot can be observed at the center of the shadow. This is due to the phenomenon of diffraction, where the waves of light bend around the edges of the object and interfere constructively at the center of the shadow[1] figure 6.11 shows a simulation of this Arago point.

In Fraunhofer diffraction, where the screen is positioned far away from the aperture, the resulting diffraction pattern is a fuzzy image that may not resemble the aperture's shape. The Fresnel diffraction pattern, however, is a somewhat distorted image of the aperture with fringed edges. The fringe pattern arises from the interference of diffracted wavefronts, which depends on the aperture's geometry and the distance between it and the observation screen. As the observation screen moves away from the aperture, the Fresnel diffraction pattern gradually transforms into the Fraunhofer diffraction pattern.

[1] An analysis of Fresnel's theory by Poisson—a strong proponent of the particle theory of light—revealed a seemingly ridiculous implication: the wave theory, according to him, would produce a bright spot in the middle of the shadow cast by a circular obstacle. This deduction presented an opportunity to discredit Fresnel's theory and was thus used as an effective argument against it. After Arago's experiment this bright spot has usually been referred to as the Arago spot.

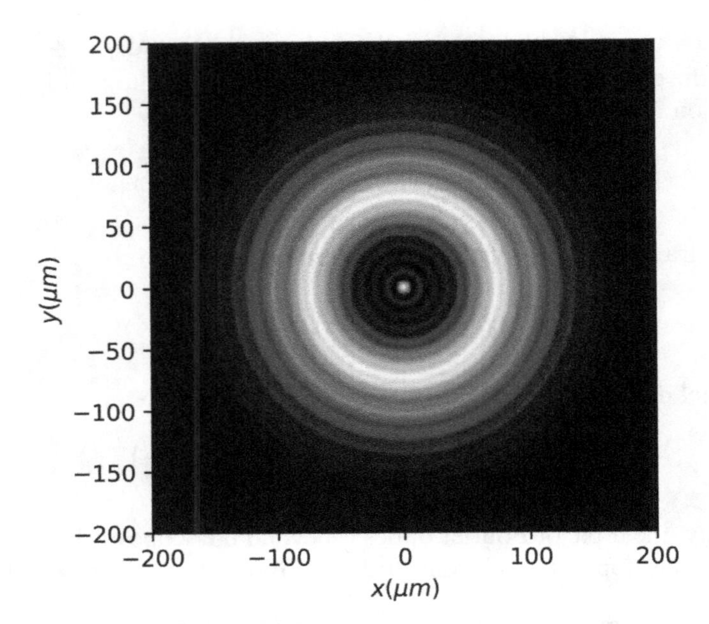

Figure 6.11. Simulation of an Arago spot.

References

[1] Born M *et al* 1999 *Principles of Optics* 7th edn (Cambridge: Cambridge University Press)

[2] Goodman J W 1996 *Introduction to Fourier Optics (Electrical Engineering Series)* (New York: McGraw-Hill)

IOP Publishing

Optical Sensors
An introduction with lab demonstrations
Victor Argueta-Diaz

Chapter 7

Optical waveguides

7.1 Introduction

Optical waveguides are structures that are designed to confine and guide the propagation of light. They are often used in sensors for highly sensitive and selective detection of analytes. This is particularly useful in applications such as medical diagnostics, environmental monitoring, and food safety testing.

The history of optical waveguides dates back to the early 20th century, when researchers first began exploring the properties of light and its behavior in various materials. However, it was not until the 1960s that the first practical optical waveguides were developed.

In 1961, Elias Snitzer, a researcher at American Optical Company, demonstrated the first optical fiber capable of guiding light over long distances. The fiber consisted of a thin glass core surrounded by a cladding layer, and was able to transmit light over several meters with relatively low attenuation.

Over the next few decades, researchers made significant advances in the field of optical waveguides. In the 1970s and 1980s, the development of low-loss fibers with improved bandwidth and reliability paved the way for the widespread adoption of fiber-optic communication systems.

In the 1980s and 1990s, researchers began exploring the use of planar waveguides for integrated optical devices, such as optical amplifiers, modulators, and filters. Planar waveguides were found to offer several advantages over fibers, including lower loss, higher integration density, and ease of fabrication.

In the 2000s and beyond, the field of optical waveguides continued to evolve, with researchers developing new types of waveguides and exploring novel applications. For example, photonic crystal waveguides were discovered in the late 1990s and have since been used to create highly efficient photonic devices with unique properties, such as photonic bandgaps.

doi:10.1088/978-0-7503-4876-8ch7

7.1.1 Design parameters

An optical waveguide typically consists of a high-index core surrounded by a lower-index cladding, which causes light to be confined to the core region by total internal reflection. There are several types of optical waveguides, each with their own unique characteristics and applications. Here are some of the most common types: planar waveguide, fiber optics, rib waveguides, and photonic crystals.

Designing an optical waveguide involves understanding the fundamental principles of waveguide optics and the specific requirements of the application. Here are some key factors to consider when designing an optical waveguide:

1. **Waveguide geometry:** the geometry of the waveguide, such as its width and thickness, can have a significant impact on its performance. The geometry should be chosen to optimize properties such as the mode confinement, propagation loss, and polarization properties.
2. **Refractive index:** the refractive index of the waveguide material is critical for determining the mode properties, such as the effective index and the mode field profile. The refractive index can be controlled through the choice of material and processing parameters.
3. **Material selection:** the choice of waveguide material is important for determining properties such as the refractive index, propagation loss, and fabrication feasibility. Materials commonly used for waveguides include silicon, silica, polymers, and glass.
4. **Fabrication techniques:** the fabrication technique used to create the waveguide can also impact its performance. Techniques such as photolithography, electron-beam lithography, and ion implantation can be used to create waveguides with high precision and reproducibility.
5. **Application-specific requirements:** the design of the waveguide should also take into account the specific requirements of the application. For example, a waveguide designed for sensing applications may need to be adapted with specific recognition molecules, while a waveguide designed for telecommunications may need to be optimized for low loss and high bandwidth.

Overall, designing an optical waveguide requires a thorough understanding of the fundamental principles of waveguide optics and the specific requirements of the application. It also requires knowledge of materials science, fabrication techniques, and device characterization methods.

In this chapter, we will present the basic principles of light propagation in optical waveguides. We will discuss the different types of optical waveguides, especially planar waveguides, rectangular waveguides, and optical fibers. We will talk about the propagation modes that can propagate in an optical waveguide, such as the fundamental mode and higher-order modes. We finally describe the various applications of optical waveguides in the area of sensor development.

7.2 Slab waveguide

We will start our description of waveguides by analyzing the slab waveguide configuration, shown in figure 7.1. This is one of the simplest configurations and will help us to understand the basic principles of light propagation in a waveguide. In our figure, we assume that the dimension in the y-direction is much larger than in the x-direction, we also assume that the core has a thickness of d. Finally, we will assume that our optical wave will be propagating in the z-direction.

A slab waveguide is a type of planar waveguide that consists of a thin rectangular slab of high-index material in between two cladding layers of lower refractive index. The slab serves as the core of the waveguide and supports the guided modes of light. If the material used in the substrate and above the core have different refractive indices, we call the structure an asymmetrical slap waveguide. If the substrate and the upper cladding have the same refractive index value, we call it a symmetric slab waveguide.

We will start our description of slab waveguides by using the Helmholtz equation for the electric field (4.12), for a wave propagating in a material with refractive index $n_i = \sqrt{\epsilon_r}$,

$$\nabla^2 \mathbf{E} + k_0^2 n_i^2 \mathbf{E} = 0 \tag{7.1}$$

where $k_0 = \omega\sqrt{\epsilon_0\mu_0}$, is the free-space wavenumber.

Due to the geometry of the slab waveguide in figure 7.1, waves will propagate in the z-direction. As a result, the z-dependence of the traveling wave will be $e^{-j\beta z}$ where β is the propagation constant in the z-direction. The magnitude of the propagation wavenumber at the core of the waveguide, is $k_0 n_{\text{core}}$ and will have components in the z- and x-direction, such that:

$$k_x^2 + \beta^2 = k_0^2 n_{\text{core}}^2 \tag{7.2}$$

The values of β and k_x are dependent not only on the values of the refractive index of the core, but also on its thickness, and wave wavelength. However, β cannot have any arbitrary value, Let's take a look at the effective wave velocity along z defined by the propagation constant β:

$$v = \frac{\omega}{\beta} = \frac{k_0}{\sqrt{\omega_0 \varepsilon_0}\,\beta} = \frac{k_0 c}{\beta} \tag{7.3}$$

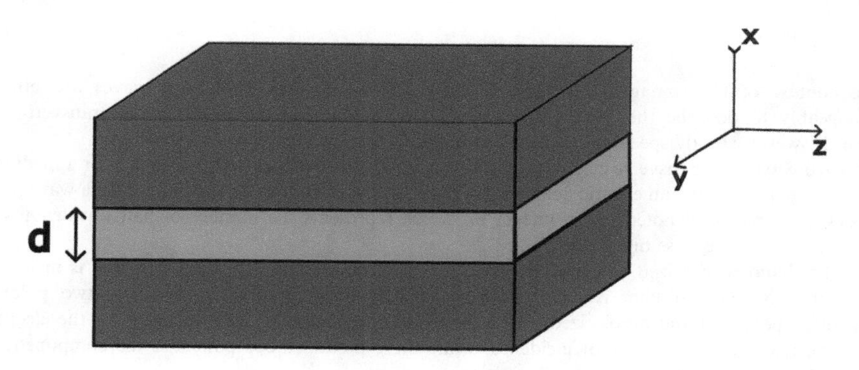

Figure 7.1. Slab waveguide configuration.

we can think of this propagation velocity as the result of an effective refractive index, $n_{\text{eff}} = \beta / k_0$. The effective refractive index, also known as the mode index or the propagation index, is a measure of the average refractive index experienced by a guided electromagnetic wave as it propagates through a waveguide. It takes into account the refractive index of the core and cladding materials, as well as the mode field distribution of the guided wave. The effective refractive index is usually greater than the refractive index of the cladding material and lower than the refractive index of the core material since the majority of the guided electromagnetic field is confined to the core. This can be expressed as:

$$k_0 n_{\text{sub}} < \beta < k_0 n_{\text{clad}} \tag{7.4}$$

We will use the Helmholtz equation to find the allowed solutions for β and $\mathbf{k}$. In the Helmholtz equation, we are assuming a uniform medium. Thus it is valid within each layer but not at the interfaces. At the interfaces, boundary conditions, (i.e. continuity of the tangential components of **E** and **H**) must be applied. Equation (7.1) can take the form:

$$\frac{\partial \mathbf{E}_0}{\partial x} + (k_0^2 n_i^2 - \beta^2)\mathbf{E}_0 = 0 \tag{7.5}$$

where we are assuming that the electric field has no variations in the y-direction. Solving this equation will give us the fields that can propagate inside our waveguide. These propagating fields are usually referred to as *propagating modes*. We have two different kinds of modes depending on the direction of the electric field, TE modes, and TM modes.

7.2.1 TE modes

TE modes, also known as transverse electric modes, are a type of propagation mode that can occur in optical waveguides. TE modes are characterized by having the electric field vector perpendicular to the plane of incidence (i.e. perpendicular to the xz-plane), while the magnetic field vector lies in the plane of incidence[1].

[1] In the context of electromagnetic waves, TE-polarized waves and s-polarized waves are often used interchangeably to describe the same thing, namely waves that are polarized in the transverse electric direction. However, strictly speaking, there is a subtle difference between the two concepts.

TE-polarized waves are transverse electromagnetic waves that are confined to a waveguide or a medium, and are characterized by having an electric field perpendicular to the direction of propagation. In a waveguide, the TE modes are typically denoted by an integer m, which represents the number of half-wavelengths of the electric field in the transverse direction.

On the other hand, s-polarized waves refer specifically to a polarization state of light that is incident on a surface. When a beam of light is incident on a surface, it can be decomposed into two polarization components: s-polarized and p-polarized. The s-polarized component is the component of the electric field that is perpendicular to the plane of incidence, while the p-polarized component is the component that is parallel to the plane of incidence.

For the waveguide described in figure 7.1, the TE wave will have its electric field in the y-direction, so equation (7.5), can be further simplified to:

$$\frac{\partial \mathbf{E}_{0y}}{\partial x} + (k_0^2 n_i^2 - \beta^2)\mathbf{E}_{0y} = 0 \tag{7.6}$$

We will take a closer at the expression in parenthesis in equations (7.6). Remember that we defined the propagation constant as $\beta = n_{\text{eff}} k_o$, and that in general, the effective refractive index is larger than the refractive index of the cladding and lower than the refractive index of the core.

So the Helmholtz equation for the core region can be rewritten as:

$$\frac{\partial \mathbf{E}_{0y}}{\partial x} + \kappa^2 \mathbf{E}_{0y} = 0, \qquad 0 \leqslant x \leqslant d \tag{7.7}$$

where:

$$\begin{aligned} \kappa &= \sqrt{k_0^2 n_{\text{core}}^2 - \beta^2} \\ &= \sqrt{k_0^2 n_{\text{core}}^2 - k_0^2 n_{\text{eff}}^2} \\ &= \sqrt{k_0^2 (n_{\text{core}}^2 - n_{\text{eff}}^2)} \end{aligned} \tag{7.8}$$

and for $n_{\text{eff}} < n_{\text{core}}$, κ will be a real number, however, for the region below the core, our Helmholtz equation will be:

$$\frac{\partial \mathbf{E}_{0y}}{\partial x} - \gamma^2 \mathbf{E}_{0y} = 0, \qquad d \leqslant x \tag{7.9}$$

where γ is an imaginary number as shown:

$$\begin{aligned} \gamma &= \sqrt{k_0^2 n_{\text{sub}}^2 - \beta^2} \\ &= \sqrt{k_0^2 n_{\text{sub}}^2 - k_0^2 n_{\text{eff}}^2} \\ &= \sqrt{k_0^2 (n_{\text{sub}}^2 - n_{\text{eff}}^2)} \\ &= j\sqrt{k_0^2 (n_{\text{eff}}^2 - n_{\text{sub}}^2)} \end{aligned} \tag{7.10}$$

Similarly, we can define an expression for the region above the core as:

$$\frac{\partial \mathbf{E}_{0y}}{\partial x} - \delta^2 \mathbf{E}_{0y} = 0, \qquad x \leqslant 0 \tag{7.11}$$

where

$$\delta = j\sqrt{k_0^2 (n_{\text{eff}}^2 - n_{\text{clad}}^2)} \tag{7.12}$$

Our electric fields need to satisfy the boundary conditions at each interface between the core and the two dielectric interfaces (at $x = 0$ and $x = d$). Basically, we need the tangential components of the electric field (i.e. E_y) and the tangential component of the magnetic field (i.e. H_z) to be continuous at each interface.

The electric field will then have the following solutions for the three regions [2, 4]:

$$\mathbf{E}_{0y} = \begin{cases} Ae^{\delta x} & x \leqslant 0 \\ B\cos(\kappa x) + C\sin(\kappa x) & 0 \leqslant x \leqslant d \\ De^{-\gamma(x-d)} & d \leqslant x \end{cases} \tag{7.13}$$

where the first and third equations represent evanescent equations, and the middle equation is an oscillatory wave with the core. Looking at continuity at $x = 0$ we can see that $A = B$, and from continuity at $x = d$, we obtain:

$$D = A\cos(\kappa d) + C\sin(\kappa d) \tag{7.14}$$

However, we will need to solve the boundary conditions for the tangential magnetic field (i.e. H_z) to find a complete solution. We will use Ampère's law from table 4.1 to find our magnetic field.

$$\mathbf{H}_{0z} = -\frac{j}{\omega\mu_0}\frac{\partial E_y}{\partial x} \tag{7.15}$$

Using this expression in equation (7.13), we obtain:

$$\mathbf{H}_{0z} = -\frac{j}{\omega\mu_0}\begin{cases} \delta Ae^{\delta x} & x \leqslant 0 \\ \kappa(-A\sin(\kappa x) + C\cos(\kappa x)) & 0 \leqslant x \leqslant d \\ \gamma(A\cos(\kappa d) + C\sin(\kappa d))e^{-\gamma(x-d)} & d \leqslant x \end{cases} \tag{7.16}$$

where we have already substituted equation (7.14) into our H-field equations. Evaluating continuity at $x = 0$, we obtain $C = \frac{\delta}{\kappa}A$, and from continuity at $x = d$ we obtain:

$$\kappa A\sin(\kappa d) - \kappa C\cos(\kappa d) = \gamma A\cos(\kappa d) + \gamma C\sin(\kappa d)$$

rearranging terms we get,

$$\kappa A\sin(\kappa d) - \gamma C\sin(\kappa d) = \gamma A\cos(\kappa d) + \kappa C\cos(\kappa d)$$

$$\sin(\kappa d)\left(\kappa - \frac{\gamma\delta}{\kappa}\right) = \cos(\kappa d)(\gamma + \delta)$$

and finally,

$$\tan(\kappa d + m\pi) = \frac{\kappa(\gamma + \delta)}{\kappa^2 - \gamma\delta} \tag{7.17}$$

In the final step, we added the term $m\pi$ to account for the periodicity of the tangent function. Depending on the integer value of m, there can be multiple solutions for the propagation constant β, which is referred to as the *mode order* β_m. Equation (7.17) contains known quantities such as the wavelength, refractive indices of each layer, and the core thickness, as well as the unknown propagation constant β. This equation can be solved numerically or graphically, for example, we can plot the left and right sides of the equation and locate their intersection point, which corresponds to a specific value of β for a TE mode propagating in our waveguide, as shown in figure 7.3.

Example 7.1. An asymmetric slab waveguide is made of a polymer layer of thickness $d = 5\ \mu\text{m}$ deposited on a silica substrate, and is covered with a protective layer. Assume a wavelength for the light source of $\lambda = 1.5\mu\text{m}$, the refractive index of the polymer to be 1.7, the refractive index of the substrate to be 1.65, and the refractive index of the protective layer to be 1.6. Find the allowed propagation modes β.

We will solve this problem both graphically and numerically. We will use Python code to reproduce figure 7.2 for the given parameters of the problem.

We can see that the intersection points are close to 0.5 μm^{-1}, 1.0 μm^{-1}, and 1.5 μm^{-1} and we find values of κ of 0.522 58 μm^{-1}, 1.037 48 μm^{-1}, and 1.5275 μm^{-1}, which corresponds to values for β of 7.101 742 16 μm^{-1}, 7.044 960 06 μm^{-1}, and 6.955 185 02 μm^{-1}, respectively.

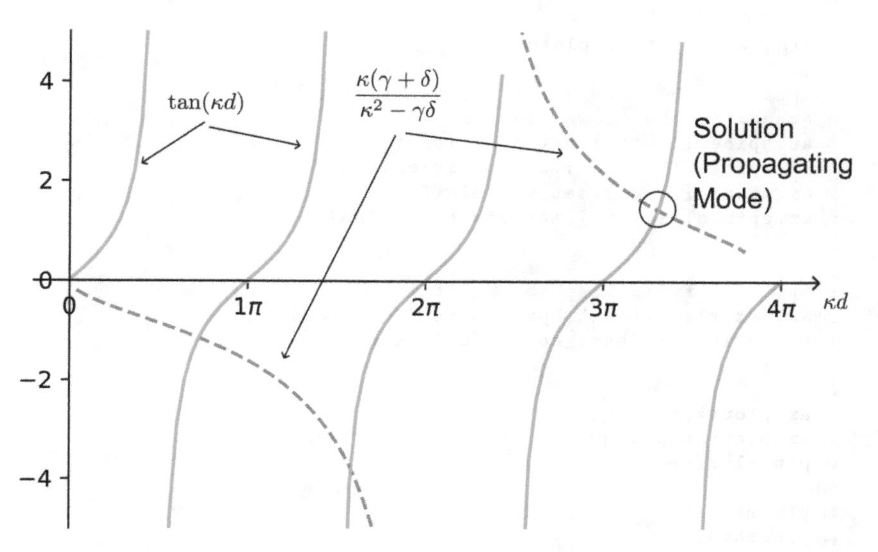

Figure 7.2. Graphical representation of equation (7.17). The intersection points represent propagating modes.

```
import numpy as np
import matplotlib.pyplot as plt


#refractive index for film, substrate, and core
nf = 1.5
ns = 1.45
nc = 1.4

#Dimensions in micrometers
h  = 5 #height of waveguide
lamb = 1 #wavelength
k = 2*np.pi/lamb


kappamax = np.sqrt(k**2*nf**2 - k**2*ns**2)
kappa = np.linspace(0, kappamax, 1000)

#propagation constants for each layer
beta = np.sqrt(k**2 *nf**2 - kappa**2)
gamma = np.sqrt(beta**2 - k**2*ns**2)
delta = np.sqrt(beta**2 - k**2*nc**2)

#calculating the dispersion curves
eq1 = np.tan(kappa*h)
eq1[:-1][np.diff(eq1) < 0 ] = np.nan #removes vertical
    lines


eq2 = kappa*(gamma+delta)/(kappa**2-gamma*delta)
eq2[:-1][abs(np.diff(eq2)) > 60 ] = np.nan



fig, ax = plt.subplots()

#placing the x-axis across the origin,
#removing the frame from the graph
ax.spines['left'].set_position(('data', 0))
ax.spines['bottom'].set_position(('data', 0))
ax.spines['top'].set_visible(False)
ax.spines['right'].set_visible(False)


#placing x label to the right
ax.set_xlabel('$\kappa$ ($\mu m^{-1}$)')
ax.xaxis.set_label_coords(0.9, 0.6)


ax.plot(kappa,eq1)
ax.plot(kappa,eq2)
plt.ylim(-5,5)

plt.savefig('GraphicalSolution.svg', format='svg')
plt.show()
```

Code 7.1:. Python code to solve graphically equation (7.17).

```
1 import numpy as np
2 from scipy.optimize import fsolve
3
4
5 #refractive index for film, substrate, and core
6 nf = 1.7
7 ns = 1.65
8 nc = 1.6
9
10 #Dimensions in micrometers
11 d  = 5 #height of waveguide
12 lamb = 1.5 #wavelength
13 k = 2*np.pi/lamb
14
15
16 #definition of our function
17 def dispersion(x):
18     kappa = x
19     beta = np.sqrt(k**2 *nf**2 - kappa**2)
20     gamma = np.sqrt(beta**2 - k**2*ns**2)
21     delta = np.sqrt(beta**2 - k**2*nc**2)
22     eq1 = np.tan(kappa*d)
23     eq2 =  kappa*(gamma+delta)/(kappa**2-gamma*delta)
24     return eq1 - eq2
25
26 #initial guesses obtained from graphical solution
27 kappa = [0.5,1.0,1.5]
28 root = fsolve(dispersion, kappa)
29 print(root)
30
31 #calculate propagation constant beta
32 print(np.sqrt(k**2 *nf**2 - root**2))
```

Code 7.2:. Python code to solve numerically equation (7.17).

With the results from our boundary conditions. We can now write an expression for our electric field at each layer

$$\mathbf{E}_{0y} = \begin{cases} Ae^{\delta x} & x \leqslant 0 \\ A\left(\cos(\kappa x) - \dfrac{\delta}{\kappa}\sin(\kappa x)\right) & 0 \leqslant x \leqslant d \\ A\left(\cos(\kappa d) + \dfrac{\delta}{\kappa}\sin(\kappa d)\right)e^{-\gamma(x-d)} & d \leqslant x \end{cases} \tag{7.18}$$

We have yet to find a value for our A coefficient to define our guided electric and magnetic fields. This coefficient is related to the power being transmitted through the waveguide. As we saw in section 4.2.2, we can calculate the power by integrating the z-component of the Poynting vector over the cross-sectional area of the waveguide.

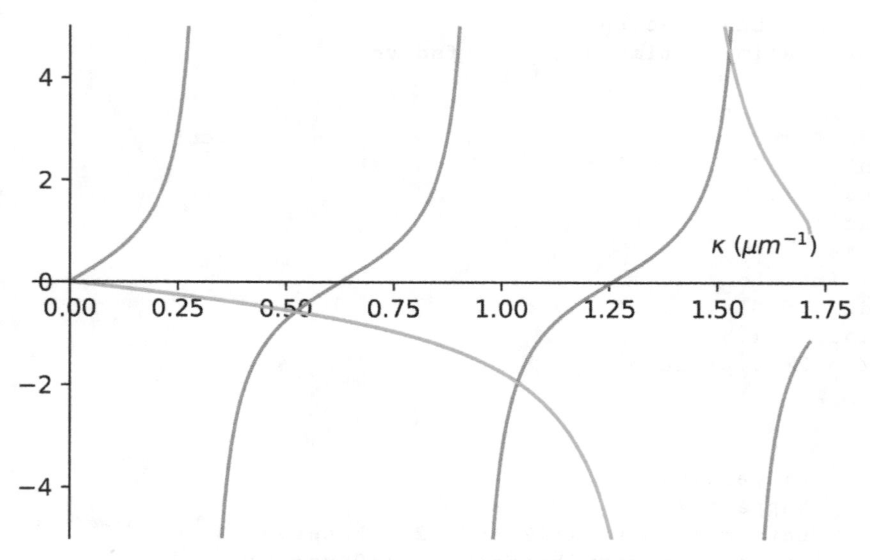

Figure 7.3. Graphical solution for the given design parameters. However, sometimes it is desired to get the numerical solutions of this equation. We will use the function *fsolve* in Python to solve this system.

$$S_z = \frac{1}{2} Re(\mathbf{E} \times \mathbf{H} \cdot z) \tag{7.19}$$

The average power for the TE mode can be found to be:

$$A = P_z = \frac{1}{2} \int_{-\infty}^{\infty} E_y H_x dx = \left(\frac{\beta}{2\omega\mu_0} \right) \int_{-\infty}^{\infty} |E_y|^2 dx \tag{7.20}$$

Example 7.2. Plot the electric field for the first three TE modes that propagate in a waveguide with the following design parameters.

Core refractive index: 1.7
Substrate refractive index: 1.4
Cladding refractive index: 1.3
Wavelength: 650 nm
Slab height: 2 μm

In order to solve this problem, we will follow the following steps:

1. Plot equation (7.17) with the design parameters of our waveguide to obtain an approximate solution for the propagation constants.
2. Solve equation (7.17) numerically, the approximate solutions obtained in the previous point will help us as initial guessed solutions.
3. Plot equations (7.18), for each layer of the waveguide.

A possible implementation to plot the *E*-field is shown in the following code.

```
1  import numpy as np
2  import matplotlib.pyplot as plt
3
4  #refractive index for film, substrate, and core
5  nf = 1.7
6  ns = 1.4
7  nc = 1.3
8
9  #Dimensions in micrometers
10 d  = 2 #height of waveguide
11 lamb = 0.65 #wavelength
12 k = 2*np.pi/lamb
13
14 i = 1
15
16 omega = 2*np.pi*3e14/lamb
17 mu = 1.2566e-12 #mu in H/um
18
19 #kappaValues = [1.346, 2.681, 3.985]
20 kappaValues = [3.985]
21
22 #define the x-values for each region
23 xsubs = np.linspace(-1,0,100)
24 xcore = np.linspace(0,d,100)
25 xclad = np.linspace(d,d+1,100)
26
27
28
29 for kappa in kappaValues:
30     #calculate propagation constants
31     beta = np.sqrt(k**2 *nf**2 - kappa**2)
32     gamma = np.sqrt(beta**2 - k**2*ns**2)
33     delta = np.sqrt(beta**2 - k**2*nc**2)
34
35     #calculation of E-fields in each region
36     fieldsubs = np.exp(delta*xsubs)
37     fieldcore = np.cos(kappa*xcore) + gamma/kappa*np.sin(
       kappa*xcore)
38     fieldclad = (np.cos(kappa*d) + gamma/kappa*np.sin(
       kappa*d))*np.exp(-gamma*(xclad-d))
39
40
41     plt.figure(i)
42     plt.plot(xsubs, fieldsubs/np.max(fieldcore), 'b')
43     plt.plot(xcore, fieldcore/np.max(fieldcore), 'orange')
44     plt.plot(xclad, fieldclad/np.max(fieldcore), 'b')
45     plt.axvline(x = 0, color = 'black', linestyle= '--')
46     plt.axvline(x = 2, color = 'black', linestyle='--')
47     plt.axhline(y=0, color ='black', linestyle = '--')
48     i+=1
49
50 #plt.savefig('Efield3.svg', format='svg')
51 plt.show()
```

Code 7.3:. Python code to plot E-fields as expressed in equation (7.18).

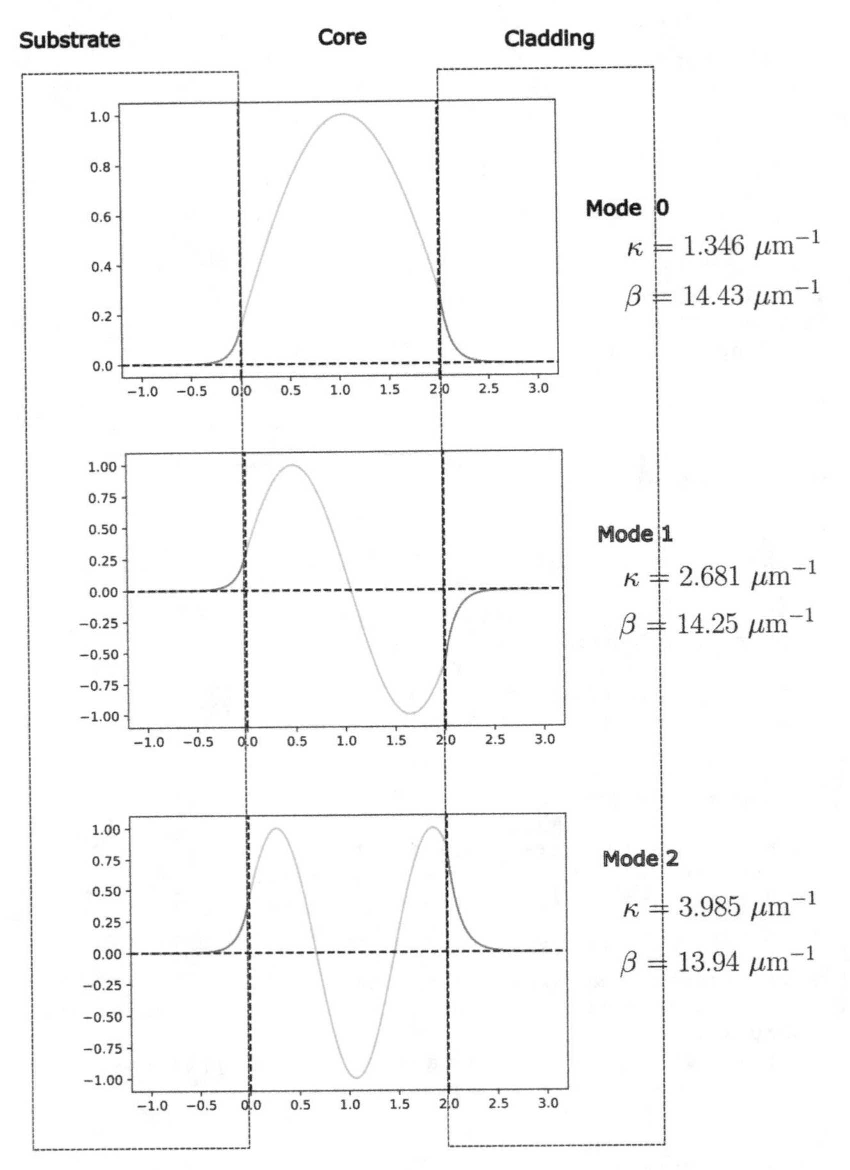

Figure 7.4. First three TE modes for the described waveguide.

The previous code produces the plots shown in figure 7.4[2].

If we look closely at the interfaces between each layer, ($x = 0$, and $x = 2$), we can see that the evanescent fields are not symmetric. The evanescent fields propagate longer on the cladding than on the substrate.

[2] In this case, we have normalized results for a better display of the figures, but the amplitude of the fields can be calculated using equation (7.20).

7.2.2 Normalized parameters

Sometimes it is convenient to normalize the design parameters of a slab waveguide. This can help to find the solutions for propagating modes for different configurations of waveguides. These normalized parameters are:

$$b = \frac{n_{\text{eff}}^2 - n_{\text{sub}}^2}{n_{\text{core}}^2 - n_{\text{sub}}^2}, \text{ normalized propagation constant} \tag{7.21}$$

$$V = k_0 d(n_{\text{core}}^2 - n_{\text{sub}}^2)^{1/2}, \text{ normalized frequency} \tag{7.22}$$

$$a = \frac{n_{\text{sub}}^2 - n_{\text{clad}}^2}{n_{\text{core}}^2 - n_{\text{sub}}^2}, \text{ asymmetric grade} \tag{7.23}$$

Because the values of the effective refractive index are in the range $n_{\text{sub}} < n_{\text{eff}} < n_{\text{clad}}$, the normalized value of b is going to be between $0 < b < 1$. The normalized frequency gives us a relationship between the thickness of the waveguide with respect of the wavelength of our wave. Finally, the asymmetric grade will be zero when we are working with a symmetric waveguide ($n_{\text{sub}} = n_{\text{clad}}$), and will increase as the difference between the substrate and cladding increases.

We can now rewrite equation (7.17) in terms of the normalized parameters as [2, 4]:

$$V\sqrt{1-b} = m\pi + \arctan\left(\sqrt{\frac{b}{1-b}}\right) + \arctan\left(\sqrt{\frac{b+a}{1-b}}\right) \tag{7.24}$$

Let's take a closer look at figure 7.5, it shows the solution of equation (7.24) for a symmetric and asymmetric waveguide as a function of their normalized parameters

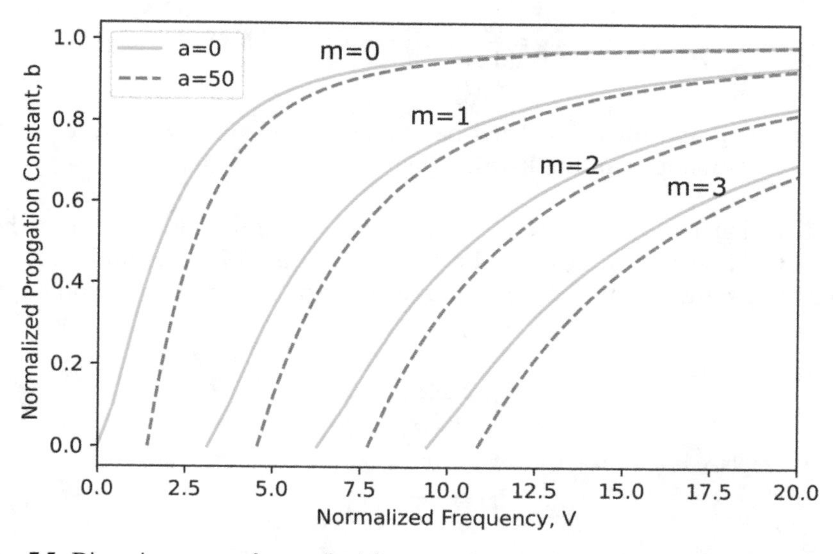

Figure 7.5. Dispersion curves of normalized frequency V versus normalized propagation constant b.

b and V. We have included the solution for the mode orders $m = 0$ to $m = 4$. For example, notice that any waveguide with a V-value lower than 2.5 will propagate only one mode. A waveguide with a V-value of 7 will propagate three modes on a symmetric waveguide ($a = 0$), but only two in an asymmetric waveguide ($a = 50$). If a waveguide allows more than one possible solution, we refer to it as a multi-mode waveguide. In the case in which there is only one possible solution, the waveguide is called a mono-mode waveguide. The guided mode with $m = 0$ is called the *fundamental mode*, and those with higher m-values are called *higher-order modes.*

Example 7.3. Let's repeat one of our previous examples, but this time we will be using normalized parameters. We need to find the propagation constants, β for an asymmetric waveguide with $n_{\text{core}} = 1.7$, $n_{\text{sub}} = 1.65$, $n_{\text{clad}} = 1.6$, $d = 5\ \mu\text{m}$ and $\lambda = 1.5\ \mu$ m.

$$\begin{aligned} a &= \frac{n_{\text{sub}}^2 - n_{\text{clad}}^2}{n_{\text{core}}^2 - n_{\text{sub}}^2} \\ &= \frac{1.65^2 - 1.6^2}{1.7^2 - 1.65^2} \\ &= 0.97 \end{aligned}$$

For a waveguide of thickness of 5 μm and a wavelength of 1.55 μm, we can find a normalized frequency, V.

$$\begin{aligned} V &= k_0 d (n_{\text{core}}^2 - n_{\text{sub}}^2)^{1/2} \\ &= \frac{2\pi d}{\lambda}(n_{\text{core}}^2 - n_{\text{sub}}^2)^{1/2} \\ &= \frac{10\pi}{1.5}\sqrt{1.7^2 - 1.65^2} \\ &= 8.57 \end{aligned}$$

From the graph for normalized parameters V versus b, we can locate the modes corresponding to our V-value (figure 7.6).

Notice that our line at $V = 8.57$ only intersects three modes. From these points of intersection, we can read three values of b: 0.2, 0.65, and 0.95.

We can use the last normalized parameter to obtain our propagation modes.

$$b = \frac{\left(\frac{\beta}{k_0}\right)^2 - n_{\text{sub}}^2}{n_{\text{core}}^2 - n_{\text{sub}}^2}$$

where we replace $n_{\text{eff}} = \beta/k_0$, solving for β:

$$\beta = k_0\sqrt{(n_{\text{core}}^2 - n_{\text{sub}}^2)b + n_{\text{sub}}^2}$$

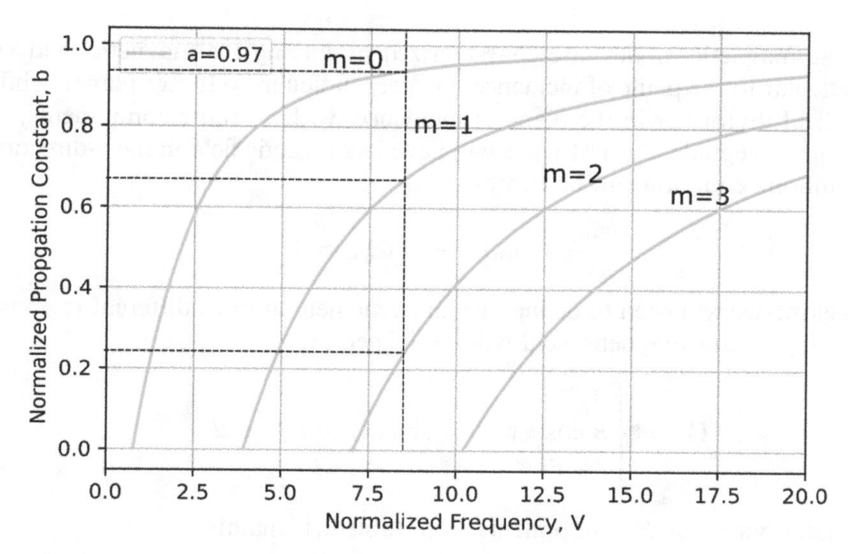

Figure 7.6. Normalized dispersion curves for given design parameters.

finally we get $\beta = 6.955\ \mu\text{m}^{-1}$, $7.04\ \mu\text{m}^{-1}$, and $7.101\ \mu\text{m}^{-1}$. Which are solutions pretty close to those obtain before.

We can also see from figure 7.5 that as the V-number increases, either by increasing the thickness of the core or by working with a smaller wavelength, the number of allowed modes increases. When the V-number is very small (e.g., $d/\lambda \ll 1$), the effective index is close to that of the cladding layer n_{sub}, as a consequence, the evanescent waves penetrate deeply into the cladding layers. Finally, as the V-number increases, the opposite happens. The wave is tightly confined to the core of the waveguide, and the evanescent wave decays rapidly in the cladding.

An important consideration to have when designing a waveguide is to know the thickness of the core to support a specific number of modes at a specific wavelength. We can use the normalized propagating vector to find the cut-off frequency for a mode. We can assume that for values of $b < 0$ there will be no propagating mode, so our cut-off condition will be $b = 0$. We can then substitute this value of b, into equation (7.24) to find the cut-off normalized frequency.

$$V_c = m\pi + \arctan(\sqrt{a}) \tag{7.25}$$

Notice that for a symmetrical waveguide, $a = 0$, V_c has a value of zero for $m = 0$. Meaning that for a symmetric waveguide, there is no cut-off frequency for the fundamental mode.

7.2.3 TM modes

Once we have completed the derivation for TE propagation modes, we can repeat a pretty similar derivation to find our TM modes. TM modes, also known as transverse magnetic modes, are characterized by having the magnetic field vector perpendicular to the plane of incidence (i.e. perpendicular to the xz-plane), while the electric field vector lies in the plane of incidence (with x- and z components).

For our waveguide, the TM wave will have its magnetic field in the y-direction, so the Helmholtz equation can be expressed as:

$$\frac{\partial \mathbf{H}_{0y}}{\partial x} + (k_0^2 n_i^2 - \beta^2)\mathbf{H}_{0y} = 0 \tag{7.26}$$

As before, we will need to define our magnetic field in three different regions, one for each layer. Our magnetic field will be defined as:

$$\mathbf{H}_{0y} = \begin{cases} Ae^{\delta x} & x \leqslant 0 \\ B\cos(\kappa x) + C\sin(\kappa x) & 0 \leqslant x \leqslant d \\ De^{-\gamma(x-d)} & d \leqslant x \end{cases} \tag{7.27}$$

We can obtain our E_z components from table 4.1, mainly:

$$E_z = \frac{j}{\omega n_i^2 \varepsilon_0} \frac{\partial H_{0y}}{\partial_x} \tag{7.28}$$

and we get:

$$\mathbf{E}_{0z} = \frac{j}{\omega \varepsilon_0} \begin{cases} \frac{\delta A}{n_{\text{clad}}^2} e^{\delta x} & x \leqslant 0 \\ \frac{\kappa}{n_{\text{core}}^2}(A\cos(\kappa x) + C\sin(\kappa x)) & 0 \leqslant x \leqslant d \\ \frac{\gamma}{n_{\text{sub}}^2}(A\cos(\kappa x) + C\sin(\kappa x))e^{-\gamma(x-d)} & d \leqslant x \end{cases} \tag{7.29}$$

In order to solve these constants, we will need to satisfy boundary conditions, that is, for H_y, and E_z components to be continuous at $x = 0$, and $x = -d$. Once the boundary conditions are solved, we will have an equivalent equation to (7.17):

$$\tan(\kappa d + m\pi) = \frac{n_{\text{core}}^2 \kappa (n_{\text{clad}}^2 \gamma + n_{\text{sub}}^2 \delta)}{n_{\text{sub}}^2 n_{\text{clad}}^2 \kappa^2 - n_{\text{core}}^2 \gamma \delta} \tag{7.30}$$

and our normalized parameters are defined as before, with the exception of the asymmetric parameter:

$$a = \left(\frac{n_{\text{core}}}{n_{\text{sub}}}\right)^4 \left(\frac{n_{\text{sub}}^2 - n_{\text{clad}}^2}{n_{\text{core}}^2 - n_{\text{sub}}^2}\right) \tag{7.31}$$

7.2.4 Optical confinement

The confinement factor refers to the fraction of the energy of an electromagnetic wave that is confined within the waveguide's core as opposed to the energy that propagates outside of the waveguide. It measures how effectively the waveguide confines the energy of the electromagnetic wave.

Mathematically, the confinement factor is defined as the ratio of the power carried by the electromagnetic wave in the waveguide core to the total power of the wave, including both the power inside and outside the waveguide's core. Using the Poything vector to calculate the power of the transmitted wave, we can calculate the confinement power as follows:

$$\frac{P_{\text{core}}}{P_{\text{total}}} = \frac{\int_0^d E_y H_x dx}{\int_{-\infty}^{\infty} E_y H_x dx} \tag{7.32}$$

We can use the normalized parameters to simplify the previous expression [3, 4]

$$\frac{P_{\text{core}}}{P_{\text{total}}} = \frac{d}{t_{\text{eff}}}\left[1 + \frac{\sqrt{b}}{V} + \frac{\sqrt{b+a}}{V(1+a)}\right] \tag{7.33}$$

where t_{eff} is called the *effective thickness* and is defined as:

$$t_{\text{eff}} = \frac{1}{\gamma} + \frac{1}{\delta} + 1 \tag{7.34}$$

7.3 Rectangular waveguides

Up until now, we have covered the properties of planar waveguides. However, in practical device applications, rectangular waveguides are more commonly utilized. The index profile $n(x, y)$ for rectangular waveguides is dependent on both the x- and y-coordinates. There are several categories of rectangular waveguides that are distinguished by the unique distribution of their index profiles. Some rectangular waveguides include: buried channel waveguides, strip-loaded waveguides, ridge waveguides, rib waveguides, and diffused waveguides. These waveguides are depicted in the figure below (figure 7.7).

1. **Ridge waveguides:** are a type of optical waveguide that has a raised, rectangular cross-sectional structure, resembling a ridge or a mesa. The ridge is typically made of a material with a higher refractive index than the surrounding cladding material, which helps to confine light within the waveguide.
2. **Rib waveguides:** are a type of optical waveguide that has a raised rectangular cross-sectional structure, resembling a rib or fin. The rib is typically made of a higher refractive index material than the surrounding cladding material, which helps to confine light within the waveguide.

   Rib waveguides are similar to ridge waveguides, but instead of having a flat top surface, they have a raised rectangular structure that extends above the cladding layer. This rib structure provides additional confinement of light within the waveguide, resulting in higher optical confinement and lower losses.

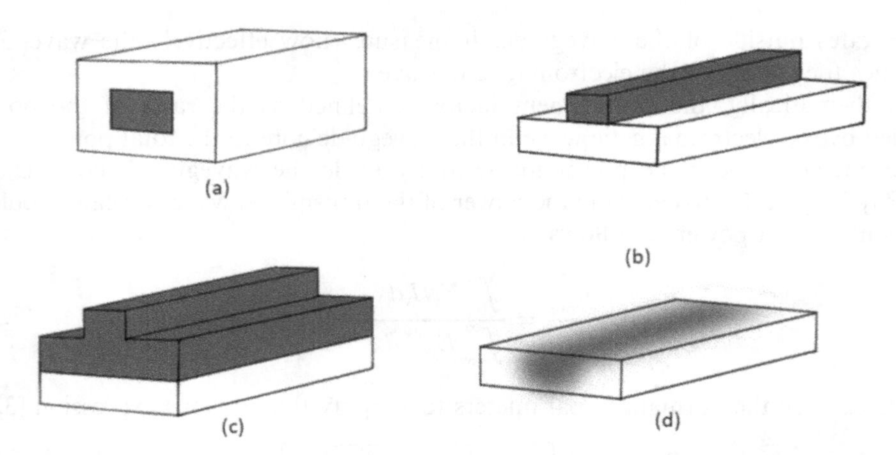

Figure 7.7. Different kinds of channel waveguides. (a) Channel waveguide; (b) ridge waveguide; (c) rib waveguide; (d) diffuse waveguide. Image created by L Argueta-Slaughter.

3. **Buried channel waveguides:** are a type of optical waveguide that is formed by introducing a high refractive index material into a lower index substrate. The high index material is typically diffused or implanted into the substrate, creating a buried channel that guides light through the structure.
4. **Diffused waveguides:** are optical waveguides in which a gradient refractive index profile is created by the diffusion of dopants or impurities into the substrate material. This process increases the refractive index in the diffused region, creating a guiding structure for light.

7.3.1 Field shadows method

A cross-section of a general structure for a rectangular waveguide is shown in figure 7.8. An exact analytical solution for this structure is impossible to obtain due to the requirements needed to satisfy all boundary conditions simultaneously. We will first treat this problem using a method proposed by Mercatili in 1969 [1]. His method works on the assumption that modes that are propagating will be well above the cut-off frequency. Therefore, these propagating modes will be well confined at the core and will decay rapidly in the shaded areas so that we can ignore the boundary conditions in those interfaces (this method is also known as 'Mercatili's Method').

Another complication when working with rectangular waveguides is that there are no pure TE or TM modes but rather hybrid TEM modes. However, these TEM modes are highly polarized along the x- or y-direction, so we tend to classify these polarizations in terms of the major component of the electric field. Polarizations with their main electric field component along the x-axis are called E^x_{pq} and behave similarly to a TM mode in a planar waveguide. These modes are called *quasi-TM modes*. The subscripts p, and q represent the number of nodes of the electric field E_x in the x- and y-direction, respectively. In a similar way, E^y_{pq} have the E_y field as their

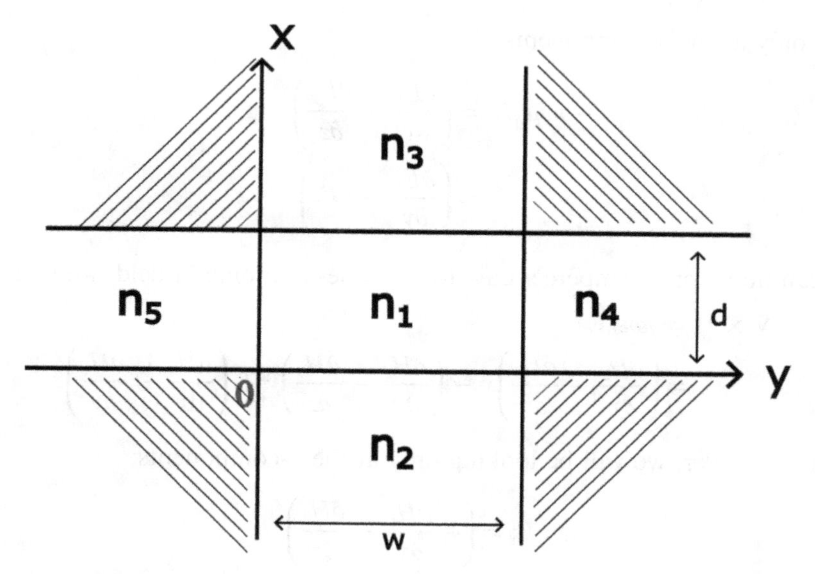

Figure 7.8. General structure for a rectangular waveguide, where the central region represents the core.

major component, they behave like a TE mode in a planar waveguide and are known as *quasi-TE modes*.

The $\mathbf{E}_0(x, y)$, and $\mathbf{H}_0(x, y)$, fields may contain transverse and longitudinal components. When solving for the Helmholtz equation we would need to solve for each of these components. We can, however, simplify the number of calculations by using the curl operations of table 4.1. First, we are going to find the derivatives of the transverse field components with respect of z:

$$\frac{\partial E_x}{\partial z} = -\beta E_x \tag{7.35}$$

$$\frac{\partial E_y}{\partial z} = -\beta E_x \tag{7.36}$$

and for the magnetic field components, we get:

$$\frac{\partial H_x}{\partial z} = -\beta H_x \tag{7.37}$$

$$\frac{\partial H_y}{\partial z} = -\beta H_x \tag{7.38}$$

Applying Faraday's Law for the time-harmonic **E** field, we get:

$$\nabla \times \mathbf{E} = -j\omega\mu_0 \mathbf{H}$$
$$= \left(\frac{\partial E_z}{\partial y} - \frac{\partial E_y}{\partial z}\right)\hat{x} - \left(\frac{\partial E_z}{\partial x} - \frac{\partial E_x}{\partial z}\right)\hat{y} + \left(\frac{\partial E_y}{\partial x} - \frac{\partial E_x}{\partial y}\right)\hat{z}$$

looking only at the x-components

$$\begin{aligned} -j\omega\mu_0 H_x &= \left(\frac{\partial E_z}{\partial y} - \frac{\partial E_y}{\partial z}\right) \\ &= \left(\frac{\partial E_z}{\partial y} + \beta E_z\right) \end{aligned} \tag{7.39}$$

We can now apply Ampère's Law to the time-harmonic **H** field, we get:

$$\begin{aligned} \nabla \times \mathbf{H} &= j\omega\varepsilon_0 \mathbf{E} \\ &= \left(\frac{\partial H_z}{\partial y} - \frac{\partial H_y}{\partial z}\right)\hat{x} - \left(\frac{\partial H_z}{\partial x} - \frac{\partial H_x}{\partial z}\right)\hat{y} + \left(\frac{\partial H_y}{\partial x} - \frac{\partial H_x}{\partial y}\right)\hat{z} \end{aligned}$$

this time, however, we will be looking only at the y-components:

$$\begin{aligned} j\omega\varepsilon_0 E_y &= \left(-\frac{\partial H_z}{\partial x} - \frac{\partial H_y}{\partial z}\right) \\ &= \left(-\frac{\partial H_z}{\partial x} - \beta H_x\right) \end{aligned} \tag{7.40}$$

combining equations (7.39) and (7.40), we can solve for H_x to get:

$$H_x = \frac{-j}{K_i^2}\left(\beta\frac{\partial H_z}{\partial x} - \omega\varepsilon_0 n_i^2 \frac{\partial E_z}{\partial y}\right) \tag{7.41}$$

where K_i is defined as:

$$K_i = \sqrt{k_0^2 n_i^2 - \beta^2} \tag{7.42}$$

and k_0 is the free-space wavevector, β is the propagation constant that we need to solve for, and n_i is the refractive index of each region. We can obtain the rest of the transversal components for **H** and **E** fields as:

$$E_x = \frac{-j}{K_i^2}\left(\beta\frac{\partial E_z}{\partial x} + \omega\mu_0\frac{\partial H_z}{\partial y}\right) \tag{7.43}$$

$$E_y = \frac{-j}{K_i^2}\left(\beta\frac{\partial E_z}{\partial y} - \omega\mu_0\frac{\partial H_z}{\partial x}\right) \tag{7.44}$$

and,

$$H_y = \frac{-j}{K_i^2}\left(\beta\frac{\partial H_z}{\partial y} + \omega\varepsilon_0 n_i^2 \frac{\partial E_z}{\partial x}\right) \tag{7.45}$$

The significance of the last five equations is that we can define the complete electromagnetic field in our waveguide based only on their longitudinal components, E_z, and H_z. We can now apply Helmholtz equation to solve for these two fields:

$$\frac{\partial^2 E_z}{\partial x^2} + \frac{\partial^2 E_z}{\partial y^2} + [k_0^2 n_i^2 - \beta^2]E_z = 0 \tag{7.46}$$

and

$$\frac{\partial^2 H_z}{\partial x^2} + \frac{\partial^2 H_z}{\partial y^2} + [k_0^2 n_i^2 - \beta^2]H_z = 0 \tag{7.47}$$

to find the solution, we will need to solve the Helmholtz equations and take into account the necessary boundary conditions.

One of the consequences of working with the above cut-off frequency is that by ignoring the boundary conditions in the shaded regions of our waveguide structure we can write our E_z field as the product of two independent functions as:

$$E_Z(x, y) = X(x)Y(y) \tag{7.48}$$

We will now find the solutions for an E^x_{pq} mode. This mode will have most of the electric field polarized in the x-direction, we can assume that $E_y \approx =0$ outside region 1, and that it will have $H_x = 0$ for all regions. For this mode for region 1, K_1 should be a real number, that is: $\beta^2 < k_0^2 n_1^2$.

The fields could be expressed as:

$$E_z = A_0 \cos(\kappa_x(x + \phi_x)) \cos(\kappa_y(y + \phi_y)) \exp(-j\beta z) \tag{7.49}$$

and for H_z we have:

$$H_z = A_1 \sin(\kappa_x(x + \phi_x)) \sin(\kappa_y(y + \phi_y)) \exp(-j\beta z) \tag{7.50}$$

where ϕ_x and ϕ_y are phase constants, A_0 and A_1 are an amplitude coefficients, and κ_x, and κ_y are propagation constants in the x- and y-direction. We will use boundary conditions to determine these unknowns.

To find C_1 we will substitute our E_z and H_z expressions into equation (7.41) and solve for $H_x = 0$, which is required for an E^x mode, we get:

$$A_1 = -A_0 n_1^2 \sqrt{\frac{\varepsilon_0}{\mu_0}} \frac{\kappa_y k_0}{\kappa_x \beta} \tag{7.51}$$

For E_z to represent a valid oscillatory function in region 1, κ_x, and κ_y must satisfy the following relationship:

$$\beta^2 = k_0^2 n_1^2 - \kappa_x^2 - \kappa_y^2 \tag{7.52}$$

Outside the core, however, the fields must have at least one component to decay exponentially; for example, in regions 2 and 3, the y-components have to be the same as the y-component in region 1, and decay in the x-direction. For regions 4 and 5, however, the x-components have to be the same as the x-component in region 1, and decay in the y-direction.

Substituting equations (7.49), and (7.50), into equation (7.43) we can obtain the E_x component for region 1. Repeating this process for all five regions[3] we obtain:

$$E_x = \begin{cases} C_1 \sin(\kappa_x(x + \phi_x)) \cos(\kappa_y(y + \phi_y)), & \text{for region 1} \\ C_2 \cos(\kappa_y(y + \phi_y)) \exp(-j\gamma_2 x), & \text{for region 2} \\ C_3 \cos(\kappa_y(y + \phi_y)) \exp(j\gamma_3(x - d)), & \text{for region 3} \\ C_4 \cos(\kappa_x(x + \phi_x)) \exp(j\gamma_4(y - w)), & \text{for region 4} \\ C_5 \cos(\kappa_x(x + \phi_x)) \exp(j\gamma_5 y), & \text{for region 5} \end{cases} \tag{7.53}$$

γ_i is a decay factor calculated by:

$$\gamma_i = \sqrt{k_0^2(n_1^2 - n_i^2) - \kappa_x^2} \tag{7.54}$$

and the amplitude coefficients C_2–C_5 are:

$$C_1 = \frac{jA_0}{\kappa_x\beta}(k_0^2 n_1^2 - \kappa_x^2) \tag{7.55}$$

$$C_2 = jA_0 \frac{\gamma_2^2 + n_2^2 k_0^2}{\gamma_2\beta} \cos(\kappa_x(d + \phi_x)) \tag{7.56}$$

$$C_3 = jA_0 \frac{\gamma_3^2 + n_3^2 k_0^2}{\gamma_3\beta} \cos(\kappa_x\phi_x) \tag{7.57}$$

$$C_4 = jA_0 \frac{n_1^2}{n_4^2} \frac{k_0^2 n_4^2 - \kappa_x^2}{\kappa_x\beta} \cos(\kappa_y(w + \phi_y)) \tag{7.58}$$

$$C_5 = jA_0 \frac{n_1^2}{n_5^2} \frac{k_0^2 n_5^2 - \kappa_x^2}{\kappa_x\beta} \cos(\kappa_y\phi_y) \tag{7.59}$$

We'll set the H_z in the four remainder regions as [2, 5]:

$$H_z = \begin{cases} C_6 \sin(\kappa_y(y + \phi_y)) \exp(-j\gamma_2 x), & \text{for region 2} \\ C_7 \sin(\kappa_y(y + \phi_y)) \exp(j\gamma_3(x - d)), & \text{for region 3} \\ C_8 \cos(\kappa_x(x + \phi_x)) \exp(j\gamma_4(y - w)), & \text{for region 4} \\ C_9 \cos(\kappa_x(x + \phi_x)) \exp(j\gamma_5 y), & \text{for region 5} \end{cases} \tag{7.60}$$

[3] For a complete description of the electric and magnetic fields in a rectangular waveguide see appendix B.

where the amplitude coefficients C_6–C_9 are:

$$C_6 = -A_0 n_2^2 \sqrt{\frac{\varepsilon_0}{\mu_0}} \frac{\kappa_y k_0}{\gamma_2 \beta} \cos(\kappa_x \phi_x) \tag{7.61}$$

$$C_7 = A_0 n_3^2 \sqrt{\frac{\varepsilon_0}{\mu_0}} \frac{\kappa_y k_0}{\gamma_3 \beta} \cos(\kappa_x (d + \phi_x)) \tag{7.62}$$

$$C_8 = -A_0 n_1^2 \sqrt{\frac{\varepsilon_0}{\mu_0}} \frac{\gamma_4 k_0}{\kappa_x \beta} \cos(\kappa_y \phi_y) \tag{7.63}$$

$$C_9 = -A_0 n_1^2 \sqrt{\frac{\varepsilon_0}{\mu_0}} \frac{\gamma_5 k_0}{\kappa_x \beta} \cos(\kappa_y (w + \phi_y)) \tag{7.64}$$

For our example, where an E^x mode is propagating. We must solve for continuity of H_z between regions 1 and 2 at $x = 0$, and between regions 1 and 3 at $x = d$. From the continuity at $x = 0$ for H_z:

$$C_6 = A_1 \sin(\kappa_x \phi_x) \tag{7.65}$$

substituting the values for A_1 and C_6

$$\begin{aligned} \frac{n_2^2}{\gamma_2} \cos(\kappa_x \phi_x) &= \frac{n_1^2}{\kappa_x} \sin(\kappa_x \phi_x) \\ \tan(\kappa_x \phi_x) &= \frac{n_2^2 \kappa_x}{\gamma_2 n_1^2} \end{aligned} \tag{7.66}$$

We can now look at the continuity of H_z between regions 1 and 3 at $x = d$:

$$C_7 = A_1 \sin(\kappa_x (x + \phi_x)) \tag{7.67}$$

substituting the values for A_1 and C_7

$$\begin{aligned} \frac{n_3^2}{\gamma_3} \cos(\kappa_x (d + \phi_x)) &= -\frac{n_1^2}{\kappa_x} \sin(\kappa_x (x + \phi_x)) \\ \tan(\kappa_x (x + \phi_x)) &= -\frac{n_1^2 \gamma_3}{n_3^2 \kappa_x} \end{aligned} \tag{7.68}$$

eliminating ϕ_x from equations (7.66) and (7.68) we obtain:

$$\tan(\kappa_x d + p\pi) = \frac{n_1^2 \kappa_x (n_3^2 \gamma_2 + n_2^2 \gamma_3)}{n_2^2 n_3^2 \kappa_x^2 - n_1^4 \gamma_2 \gamma_3} \tag{7.69}$$

which can be recognized as equation (7.30), which we derived for TM modes in a slab waveguide.

We can repeat the same process to look at the continuity of H_z between regions 1 and 5 at $y = 0$, and between regions 1 and 4 at $y = w$, we can get the following characteristic equations for κ_y:

$$\tan(\kappa_y \phi_y) - \frac{\gamma_4}{\kappa_y} \tag{7.70}$$

$$\tan(\kappa_y(w + \phi_y)) = \frac{\gamma_5}{\kappa_y} \tag{7.71}$$

eliminating ϕ_y from the previous two equations, we obtain:

$$\tan(\kappa_y w + q\pi) = \frac{\kappa_y(\gamma_4 + \gamma_5)}{\kappa_y^2 - \gamma_4\gamma_5} \tag{7.72}$$

Which is the same as equation (7.17), that we derived for TE modes in a slab waveguide. Once we determined the values of κ_x and κ_y, either by graphical or numerical methods as we did previously, we can use equation (7.52) to find the value of the propagation constant β.

7.3.2 Normalized parameters

This method simplifies the analysis of waveguide structures and allows the calculation of propagation constants and mode profiles for different waveguide geometries and materials.

We start our problem by splitting our rectangular waveguide into two slab waveguides, as shown in figure 7.9. One slab waveguide will be along the x-axis, and the other one will be along the y-axis. In order to keep the right values of refractive index in each region, we will modify the values of the refractive indices of the slabs to $n_c^2/2$ for the core and $n_s^2 - n_c^2/2$. for the substrate and cladding.

We can solve the propagation constant and effective refractive index for each slab section by repeating the process already described in section 7.2.

The propagation constant of the rectangular waveguide will be then:

$$\beta^2 = \beta_x^2 + \beta_y^2 \tag{7.73}$$

and the effective refractive index:

$$n_{\text{eff}}^2 = n_{\text{eff}x}^2 + n_{\text{eff}y}^2 \tag{7.74}$$

We can also find the normalized parameters of the rectangular waveguide based on the normalized parameters of the slab components as:

$$V_x = kd\sqrt{n_{\text{core}}^2 - n_{\text{subs}}^2} = V \tag{7.75}$$

$$V_y = kw\sqrt{n_{\text{core}}^2 - n_{\text{subs}}^2} = \frac{w}{d}V \tag{7.76}$$

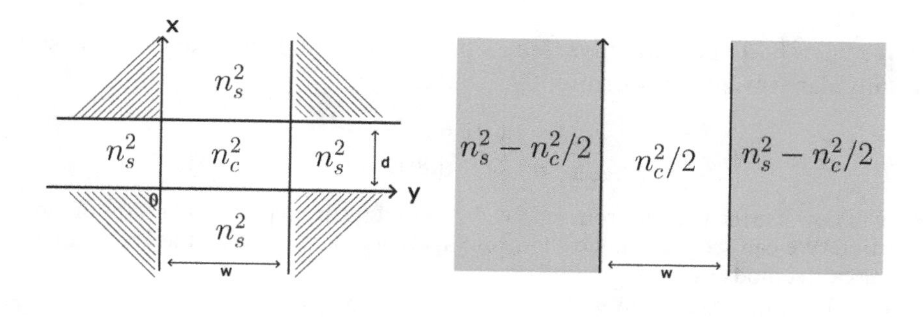

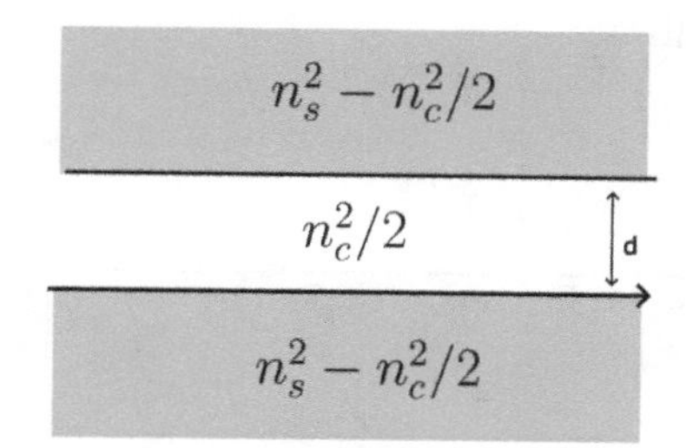

Figure 7.9. Splitting a rectangular waveguide into two slab waveguides.

and our normalized propagation vector can be calculated from:

$$b_x = \frac{n_{\text{effx}}^2 - n_{\text{subs}}^2 + n_{\text{core}}^2/2}{n_{\text{core}}^2 - n_{\text{subs}}^2} \tag{7.77}$$

$$b_y = \frac{n_{\text{effy}}^2 - n_{\text{subs}}^2 + n_{\text{core}}^2/2}{n_{\text{core}}^2 - n_{\text{subs}}^2} \tag{7.78}$$

and, for the rectangular waveguide:

$$b = \frac{n_{\text{eff}}^2 - n_{\text{subs}}^2}{n_{\text{core}}^2 - n_{\text{subs}}^2} \tag{7.79}$$

we can also obtain b from the following relationship:

$$b = b_x - b_y - 1 \tag{7.80}$$

With these new normalized parameters, we can use figure 7.5 to determine the normalized guided index for a rectangular waveguide.

Example 7.4. Find the fundamental propagation constant for a rectangular waveguide with $n_{\text{core}} = 1.8$ surrounded by a material with $n_s = 1.6$, assume a rectangular core of 2 μm, and a wavelength $\lambda = 650$ nm.

We first find the refractive index of the core and cladding of the split rectangular waveguide horizontally. That is:

$$n_{\text{core}} = 1.8^2/2 = 1.62$$
$$n_{\text{cladding}} = 1.6^2 - 1.8^2/2 = 0.94$$

We will now use the process from section 7.2 to find the propagation constant in the x-direction. We can see that we have five propagating modes by using the graphical and numerical methods as before.

The fundamental mode will be $\beta_x = 9.006\ 8652\ \mu\text{m}^{-1}$.

By symmetry β_y will have the same value, and therefore, the propagation constant for our rectangular waveguide will be:

$$\beta^2 = \beta_x^2 + \beta_y^2$$

which gives us a propagation constant $\beta = 12.737\ 631\ \mu\text{m}^{-1}$.

7.4 Optical fibers

Optical fibers were first developed in the 1960s by researchers at Corning Glass Works and Bell Laboratories in the United States. In 1970, Corning Glass Works produced the first low-loss optical fiber, which had a loss of 17 dB km^{-1} at 633 nm wavelength. This breakthrough paved the way for the commercialization of optical fibers for telecommunications and other applications. Optical fibers are commonly used in optical sensors because they transmit light over long distances with very low signal loss. Optical sensors use the interaction of light with the sensing element to measure a physical quantity, such as temperature, pressure, strain, or chemical concentration.

The sensing element can be attached to or integrated within the optical fiber, allowing the light signal to interact with the sensing element and provide information about the physical quantity being measured. The change in the optical signal, such as intensity, wavelength, or polarization, is then detected and analyzed to determine the value of the physical quantity being measured.

7.4.1 Maxwell equations in cylindrical coordinates

Optical fibers are similar to rectangular waveguides in that they depend on total internal reflections from the boundary of a high refractive index core embedded in a lower refractive index cladding. Optical fibers support different types of modes, such as TE and TM, but unlike slab waveguides, but they also include hybrid modes, including a fundamental one that is a specific hybrid mode with variations in electric and magnetic fields. To understand the full range of modes that exist in optical fibers, we need to examine the solutions of Maxwell's wave equation in cylindrical coordinates. Figure 7.10, shows the basic structure of a step-index optical fiber.

We will need to solve the Helmholtz equation in cylindrical coordinates, to do so our electric field will take the following expression:

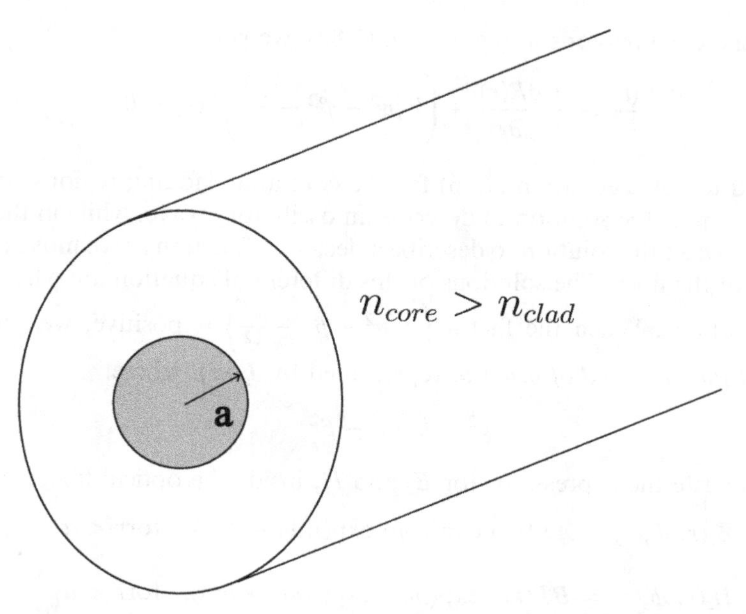

Figure 7.10. Optical fiber, consisting of a high refractive index core of radius a, and lower refractive index cladding.

$$\mathbf{E}(r, \phi, z) = \hat{r}E_r(r, \phi, r) + \hat{\phi}E_\phi(r, \phi, r) + \hat{z}E_z(r, \phi, r) \tag{7.81}$$

One problem when solving Helmholtz equation in cylindrical coordinates is that the E_r and E_ϕ are highly coupled, however, the E_z component doesn't couple to the other two [2, 4], so it's possible to write the Helmholtz equation only in terms of E_z:

$$\frac{1}{r}\frac{\partial}{\partial r}\left(r\frac{\partial E_z}{\partial r}\right) + \frac{1}{r^2}\frac{\partial^2 E_z}{\partial \phi^2} + \frac{\partial^2 E_z}{\partial z^2} + k_0^2 n^2 E_z = 0 \tag{7.82}$$

Once we have found a solution for E_z we can use Maxwell's equations to find the E_r and E_ϕ components. Let's express $E_z(r, \phi, z)$ as a function of three separable variables such that:

$$E_z(r, \phi, z) = R(r)P(\phi)Z(z) \tag{7.83}$$

Because of the cylindrical symmetry, each field component must remain the same when ϕ increases by 2π; we thus can assume a periodic function $P(\phi)$ to be:

$$P(\phi) = \exp(j\nu\phi) \tag{7.84}$$

where ν must be a negative or positive integer. Also, the $Z(z)$ function will be our propagation along the z-axis:

$$Z(z) = \exp(-j\beta z) \tag{7.85}$$

replacing these expressions into equation (7.82), we get

$$\frac{\partial^2 R(r)}{\partial r^2} + \frac{1}{r}\frac{\partial R(r)}{\partial r} + \left(k_0^2 n^2 - \beta^2 - \frac{\nu^2}{r^2}\right) R(r) = 0 \tag{7.86}$$

We need to solve equation (7.86) for the core and cladding regions. In the core region, we expect the solution to describe an oscillatory wave, while in the cladding region, we expect the solution to describe a decaying function as we move away from the center of the fiber. The solutions of this differential equation are what are called Bessel Functions. When the factor $\left(k_0^2 n^2 - \beta^2 - \frac{\nu^2}{r^2}\right)$ is positive, we get a *Bessel function of the first kind of order* ν, represented by $J_\nu(\kappa r)$, where:

$$\kappa^2 = k_0^2 n_1^2 - \beta^2 \tag{7.87}$$

We can write the expressions for E_z and H_z inside the optical fiber's core as:

$$E_z(r, \phi, z) = AJ_v(\kappa r)\exp(j\nu\phi)\exp(j(\omega t - \beta z)), \quad \text{for} r < a \tag{7.88}$$

$$H_z(r, \phi, z) = BJ_v(\kappa r)\exp(j\nu\phi)\exp(j(\omega t - \beta z)), \quad \text{for} r < a \tag{7.89}$$

where A and B, are amplitude constants that will be defined once we apply a boundary condition between the core and the cladding. Figure 7.11 shows the plot of the first orders of the Bessel function of the first kind. We can see that each order has an oscillatory behavior that slowly decreases away from the core. The 0-order function has a finite amplitude of 1 at $r = 0$, while the rest of the modes will have an amplitude of zero at that same location.

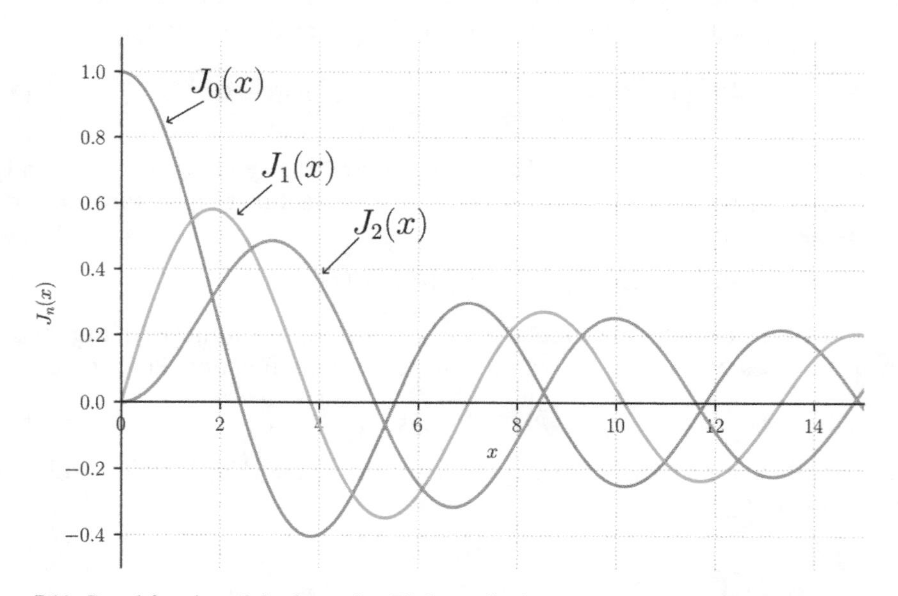

Figure 7.11. Bessel function of the first order. Their amplitude decreases as we move away from the core.

For the solution of equation (7.86) in the cladding we will expect a function that rapidly decays away from the center of the fiber. When the factor, $\left(k_0^2 n^2 - \beta^2 - \frac{\nu^2}{r^2}\right)$, is negative, the solution of this differential equation will be a *modified Bessel function of the second kind of order* ν, represented by $K_\nu \gamma r$, where:

$$\gamma^2 = \beta^2 - k_0^2 n^2 \tag{7.90}$$

we can write our E_z and H_z functions for the cladding region as:

$$E_z(r, \phi, z) = CK_\nu(\gamma r) \exp(j\nu\phi) \exp(j(\omega t - \beta z)), \quad \text{for } r > a \tag{7.91}$$

$$H_z(r, \phi, z) = DK_\nu(\gamma r) \exp(j\nu\phi) \exp(j(\omega t - \beta z)), \quad \text{for } r > a \tag{7.92}$$

where C and D are again constants to be defined by boundary conditions. Figure 7.12 shows the modified Bessel function of the second kind for the first three modes. We can see that $K_\nu(\gamma r)$ presents a clearly decreasing behavior. Higher modes decrease slower, but eventually $K_\nu(\gamma r) \to 0$ as $r \to \infty$.

7.4.2 Boundary conditions for optical fibers

The solutions for the propagation constant β, and the amplitude coefficients A, B, C, and D need to be determined by applying boundary conditions. Our boundary conditions require the tangential components of the electric field, E_ϕ and E_z, to be continuous at $r = a$. Similarly, the tangential components of the magnetic field, H_ϕ and H_z need to be continuous across the interface at $r = a$.

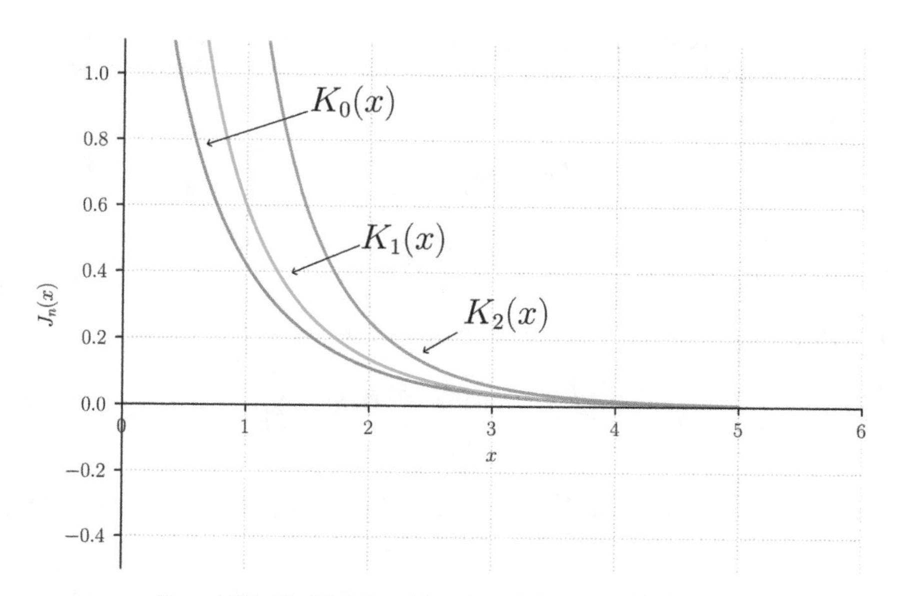

Figure 7.12. Modified Bessel function of the second kind, $K_\nu(x)$.

In order to find our E_ϕ, and H_ϕ components we will use a similar process to the one that we use to find the tangential components from the longitudinal components in a rectangular waveguide in section 7.3.1. Applying Faraday's Law and Ampère's Law for the E_z and H_z, and after some mathematical manipulation we get:

$$E_\phi = \frac{-j}{K^2}\left(\frac{\beta}{r}\frac{\partial E_z}{\partial \phi} - \omega\mu\frac{\partial H_z}{\partial r}\right) \tag{7.93}$$

$$E_\mathrm{r} = \frac{-j}{K^2}\left(\frac{\omega\mu}{r}\frac{\partial H_z}{\partial \phi} + \beta\frac{\partial E_z}{\partial r}\right) \tag{7.94}$$

$$H_\phi = \frac{-j}{K^2}\left(\omega\varepsilon\frac{\partial E_z}{\partial r} + \frac{\beta}{r}\frac{\partial H_z}{\partial \phi}\right) \tag{7.95}$$

$$H_\mathrm{r} = \frac{-j}{K^2}\left(\beta\frac{\partial H_z}{\partial r} - \frac{\omega\varepsilon}{r}\frac{\partial E_z}{\partial \phi}\right) \tag{7.96}$$

where K^2 can be either κ^2 for the core region or γ^2 for the cladding region. We will use equations (7.93) and (7.95) to get two of our tangential fields. For $r < a$ we get:

$$E_\phi = \frac{-j\beta}{\kappa^2}\left(\frac{j\nu}{r}AJ_\nu(\kappa r) + \frac{\omega\mu}{\beta}B\kappa J_\nu'(\kappa r)\right) \tag{7.97}$$

$$H_\phi = \frac{-j\beta}{\kappa^2}\left(\frac{j\nu}{r}BJ_\nu(\kappa r) - \frac{\omega\varepsilon_1}{\beta}A\kappa J_\nu'(\kappa r)\right) \tag{7.98}$$

each field also has an $\exp(j\nu\phi)\exp(-j\beta z)$ term that we do not include in the previous expressions for simplicity. n_1 is the refractive index of the core, finally we also define $J_n'u(x)$ and $K_n'u(x)$ as:

$$J_n'u(x) = \frac{dJ_\nu(x)}{dx} \tag{7.99}$$

$$K_n'u(x) = \frac{dK_\nu(x)}{dx} \tag{7.100}$$

The E_ϕ and H_ϕ fields for $r > a$ will be defined as:

$$E_\phi = \frac{-j\beta}{\gamma^2}\left(\frac{j\nu}{r}CK_\nu(\gamma r) + \frac{\omega\mu}{\beta}D\gamma K_\nu'(\gamma r)\right) \tag{7.101}$$

$$H_\phi = \frac{-j\beta}{\gamma^2}\left(\frac{j\nu}{r}DK_\nu(\gamma r) - \frac{\omega\varepsilon_2}{\beta}A\gamma K_\nu'(\gamma r)\right) \tag{7.102}$$

where n_2 is the refractive index of the cladding. We won't be using the E_r and H_r for our boundary conditions, so we won't be calculating them at this moment[4]. We know need to write down our boundary conditions; we will have one set of equations for E_z, H_z, E_ϕ, and H_ϕ, which implies solving a system of four simultaneous equations. For example, if we look at the boundary conditions at $r = a$, for E_z we get:

$$AJ_v(\kappa a)\exp(j\nu\phi)\exp(-j\beta z) = CK_\nu(\gamma a)\exp(j\nu\phi)\exp(-j\beta z) \tag{7.103}$$

which simplifies to:

$$AJ_v(\kappa a) - CK_\nu(\gamma a) = 0 \tag{7.104}$$

Similarly, for the continuity condition for H_z we get:

$$BJ_v(\kappa a) - DK_\nu(\gamma a) = 0 \tag{7.105}$$

and for E_ϕ we have:

$$A\frac{\beta\nu}{a\kappa^2}J_\nu(\kappa a) + B\frac{j\omega\mu}{\kappa}\kappa J_\nu'(\kappa a) + C\frac{\beta\nu}{a\gamma^2}K_\nu(\gamma a) + \frac{jD\omega\mu}{\gamma}K_\nu'(\gamma a) = 0 \tag{7.106}$$

and finally, for H_ϕ we have the following equation:

$$-A\frac{j\omega\varepsilon_1}{\kappa}J_\nu'(\kappa a) + B\frac{\beta\nu}{a\kappa^2}J_\nu(\kappa a) - C\frac{j\omega\varepsilon_2}{\gamma}K_\nu'(\gamma a) + D\frac{\beta\nu}{a\gamma^2}K_\nu(\gamma a) = 0 \tag{7.107}$$

Equations (7.104)–(7.107) can be expressed as a matrix, whose determinant needs to be set to zero to find the solution of the system.

$$\begin{bmatrix} J_v(\kappa a) & 0 & -K_\nu(\gamma a) & 0 \\ 0 & J_v(\kappa a) & 0 & -K_\nu(\gamma a) \\ \frac{\beta\nu}{a\kappa^2}J_\nu(\kappa a) & \frac{j\omega\mu}{\kappa}\kappa J_\nu'(\kappa a) & \frac{\beta\nu}{a\gamma^2}K_\nu(\gamma a) & \frac{j\omega\mu}{\gamma}K_\nu'(\gamma a) \\ -\frac{j\omega\varepsilon_1}{\kappa}J_\nu'(\kappa a) & \frac{\beta\nu}{a\kappa^2}J_\nu(\kappa a) & -\frac{j\omega\varepsilon_2}{\gamma}K_\nu'(\gamma a) & \frac{\beta\nu}{a\gamma^2}K_\nu(\gamma a) \end{bmatrix}\begin{bmatrix} A \\ B \\ C \\ D \end{bmatrix} = 0 \tag{7.108}$$

Calculating the determinant of the matrix gives us the characteristic equation for a step-index optical fiber:

$$\begin{aligned}\frac{\beta^2\nu^2}{a^2}\left[\frac{1}{\gamma^2} + \frac{1}{\kappa^2}\right]^2 = &\left(\frac{J_v'(\kappa a)}{\kappa J_v(\kappa a)} + \frac{K_v'(\gamma a)}{\gamma K_v(\gamma a)}\right) \\ &\left(k_0^2 n_1^2\frac{J_v'(\kappa a)}{\kappa J_v(\kappa a)} + k_0^2 n_2^2\frac{K_v'(\gamma a)}{\gamma K_v(\gamma a)}\right)\end{aligned} \tag{7.109}$$

To solve this equation, we will need graphic or numerical methods. As in the case of the slab and rectangular waveguide, there is a discrete number of allowed values

[4] A complete set of magnetic and electric fields is shown in appendix B.

for β. Once we have found our propagation constants, we can use our boundary condition equation again to find the amplitude coefficients A–D, it is sometimes useful to express the amplitude coefficients in terms of ratios as follows:

$$\frac{A}{C} = \frac{K_\nu(\gamma a)}{J_\upsilon(\kappa a)} \tag{7.110}$$

$$\frac{B}{D} = \frac{K_\nu(\gamma a)}{J_\upsilon(\kappa a)} \tag{7.111}$$

and finally,

$$\frac{A}{B} = \frac{j\nu\beta}{\omega\mu a}\left[\frac{1}{\kappa^2} + \frac{1}{\gamma^2}\right]\left[\frac{J_\nu'(\kappa a)}{\kappa J_\nu(\kappa a)} + \frac{K_\nu'(\gamma a)}{\gamma K_\nu(\gamma a)}\right]^{-1} \tag{7.112}$$

The ratio between A and B will be useful later on, because it shows us the relative amount of E_z and H_z, which we will use to determine the type of mode that is being propagated. Before we show an example of finding the allowed values of β, it is necessary to explain a little bit more about the propagation modes in an optical fiber.

7.4.3 Propagation modes

Let's take a close look at figure 7.11, we can see that for each order, ν, the function $J_\nu(x)$, oscillates and crosses the x-axis at different positions. These are the roots, m, of the function. Each root, is identified as $B_{\nu m}$. For a circular waveguide, there are four types of modes: TE $_{\nu m}$, TM $_{\nu m}$ EH $_{\nu m}$, and HE $_{\nu m}$.

In general, all modes in an optical fiber are hybrid modes, either EH or HE. Except, when $\nu = 0$, in which case, the left side of equation (7.109) is zero, and the equation simplifies to:

$$\left(\frac{J_\upsilon'(\kappa a)}{\kappa J_\upsilon(\kappa a)} + \frac{K_\upsilon'(\gamma a)}{\gamma J_\upsilon(\gamma a)}\right)\left(k_0^2 n_1^2 \frac{J_\upsilon'(\kappa a)}{\kappa J_\upsilon(\kappa a)} + k_0^2 n_2^2 \frac{K_\upsilon'(\gamma a)}{\gamma J_\upsilon(\gamma a)}\right) = 0 \tag{7.113}$$

Either term in equation (7.113) can be set to zero to solve this equation. If the first term is equal to zero, then our amplitude coefficient A will be zero, which implies that $E_z = 0$, and we have a TE mode. Similarly, if the second term is equal to zero, then B will be zero and $H_z = 0$ which implies a TM mode.

We can use the following Bessel identities to solve for either term:

$$\frac{J_\nu'}{\kappa J_\nu} = \pm\frac{J_{\nu\mp1}}{\kappa J_\nu} \mp \frac{\nu}{\kappa^2} \tag{7.114}$$

$$\frac{K_\nu'}{\gamma K_\nu} = \mp\frac{K_{\nu\pm1}}{\gamma K_\nu} \mp \frac{\nu}{\gamma^2} \tag{7.115}$$

for $\nu = 0$, the first term of equation (7.113), simplifies to:

$$\frac{J_1(\kappa a)}{\kappa J_0(\kappa a)} + \frac{K_1(\gamma a)}{\gamma K_0(\gamma a)} = 0 \tag{7.116}$$

which corresponds to the eigenvalue equation for TE modes. A way to solve this equation graphically is by plotting both terms of equation (7.116). The allowed values will be the intersection points. Similarly, setting $\nu = 0$, but solving for the second term instead, we get:

$$\frac{k_0^2 n_1^2 J_1(\kappa a)}{\kappa J_0(\kappa a)} + \frac{k_0^2 n_2^2 K_1(\gamma a)}{\gamma K_0(\gamma a)} = 0 \tag{7.117}$$

which corresponds to the eigenvalue equation for TM modes.

When $\nu \geqslant 1$, numerical methods are necessary to solve equation (7.109). The propagation modes will have E_z and H_z components. Depending on the relative magnitude of E_z versus H_z we will have an HE or EH mode. If E_z is larger than H_z we will have an HE mode. If H_z is larger than E_z we will have an EH mode. However, if we assume that the difference between the refractive index of the core and the refractive index of the cladding is very small, that is $n_{\text{core}} \approx n_{\text{clad}}$, we can simplify equation (7.109) to:

$$\frac{\nu^2}{a^2}\left[\frac{1}{\gamma^2} + \frac{1}{\kappa^2}\right]^2 = \left(\frac{J_v'(\kappa a)}{\kappa J_v(\kappa a)} + \frac{K_v'(\gamma a)}{\gamma J_v(\gamma a)}\right) \tag{7.118}$$

This is called the weakly guiding approximation; under this approximation, the propagation constants of several modes are very close to each other. These modes will travel at very similar propagation velocities, and it is possible to create stable superposition of some modes. So we can group these propagation modes into a single one called *linearly polarized modes*. Most of these LP modes have a very small E_z component that can usually be ignored.

The possible combinations for LP modes are:

$$\begin{aligned} &LP_{0m} \rightarrow HE_{1m} \\ &LP_{1m} \rightarrow TE_{0m} + TM_{0m} + HE_{2m} \\ &LP_{vm} \rightarrow HE_{\nu+1m} + EH_{\nu-1m} \end{aligned}$$

One advantage of LP modes is that they provide an easier way to visualize. Visualizing LP modes can help us understand the spatial distribution of the electromagnetic field in the transverse plane of an optical system. This can help us understand how the beam propagates and how it interacts with other optical components in the system. In figure 7.13 we show the power distribution for the LP_{01}, LP_{11}, LP_{11} and LP02 modes

One of the most important LP modes is the LP_{01} mode, equivalent to an HE_{11} mode. The LP_{01} is the fundamental mode, that is, the lowest-order mode that can propagate in an optical fiber. It has a fundamental Gaussian intensity profile and a circular symmetric transverse profile. In optical fibers, the LP_{01} mode is the most commonly used mode for transmitting light signals because it has the lowest attenuation and dispersion.

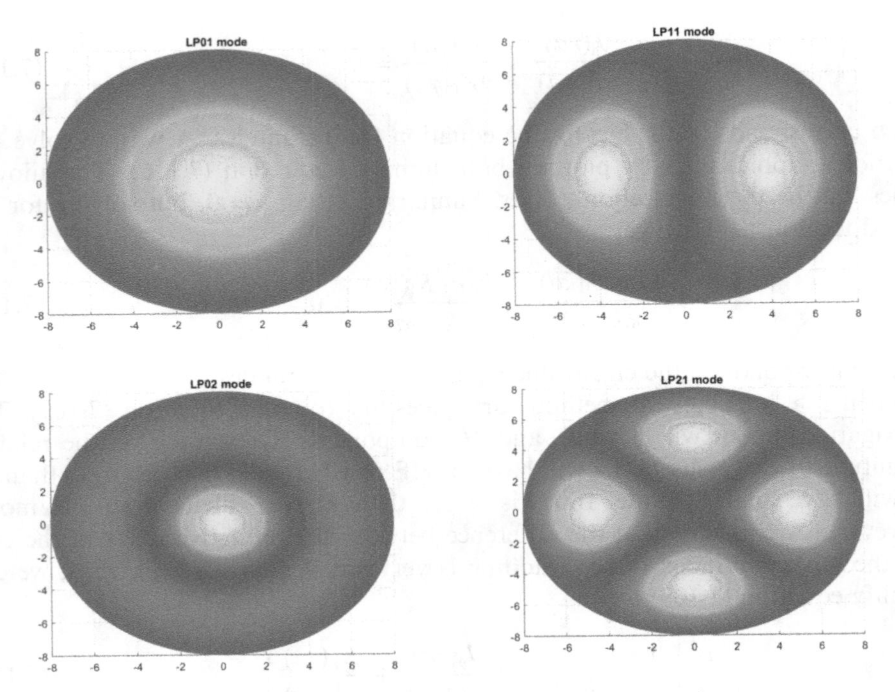

Figure 7.13. Power distribution for the first four LP modes.

7.4.4 Normalized parameters

As in the case of the slab waveguide, we can describe normalized parameters for on optical fiber. These normalized parameters are used to describe the properties of the fiber. There are different parameters that can be normalized. One of the most commons is the *Normalized frequency* (V). It is defined as:

$$V = \frac{2\pi a}{\lambda}\sqrt{n_{\text{core}}^2 - n_{\text{clad}}^2} \tag{7.119}$$

where a is the core radius, and λ is the operational wavelength. The normalized frequency is used to determine whether an optical fiber is operating in the single-mode or multi-mode regime. It can also help us determine the number of propagating modes.

The V-number is a key parameter for determining whether an optical fiber is operating in the single-mode or multi-mode regime. For single-mode fibers, the V-number is typically less than 2.405, which is known as the mode cut-off value. At V-numbers below the mode cut-off value, the fiber supports only a single mode, while at V-numbers above the mode cut-off value, the fiber supports multiple modes. The exact value of the mode cut-off V-number depends on the details of the fiber's geometry and refractive index profile. To understand where this cut-off V-number comes from, we need to talk about the normalized propagation constant. As in the case of the slab waveguide, we can also have a *normalized propagation constant*, defined by [2]:

$$b = \frac{a^2\gamma^2}{V^2} = \frac{(\beta/k)^2 - n_2^2}{n_1^2 - n_2^2} \tag{7.120}$$

Figure 7.14 shows that for $V < 2.405$, LP_{01} is the only mode propagating. LP_{01} always exists, and it has no cut-off frequency. The number of guided modes supported by the fiber increases as V increases. When a mode starts propagating the effective refractive index will be close to the cladding refractive index, so the wave will extend into the cladding. As the effective refractive index moves closer to the refractive index of the core the wave will be more confined to the core. The cut-off frequency V_c of any given mode is defined by $b(V) = 0$ and can be found by:

$$J_{\nu-1}(V_c) = 0 \tag{7.121}$$

The cut-off frequency of the $LP_{\nu m}$ mode is the mth zero of $J_{\nu-1}$. In table 7.1 we show several cut-off V-numbers for some $LP_{\nu m}$-modes.

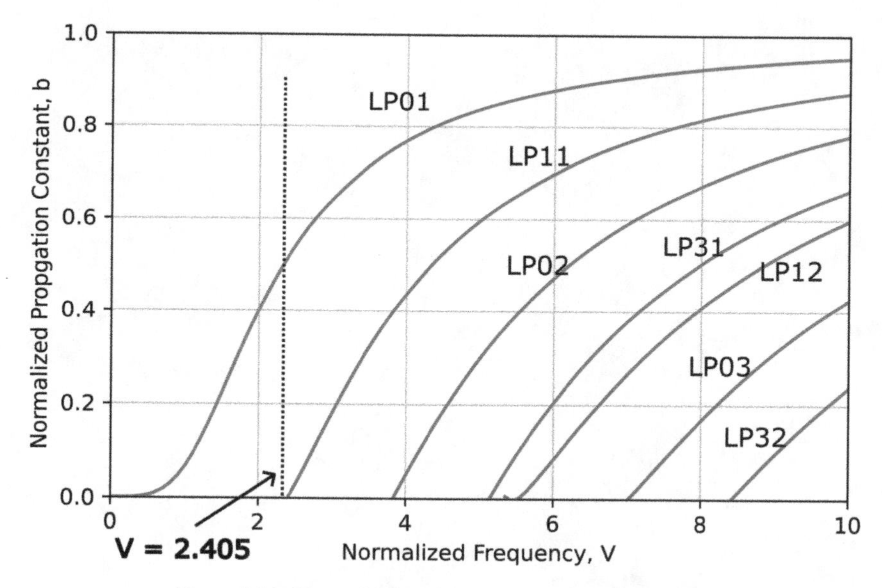

Figure 7.14. Normalized V–b curve for the first LP modes.

Table 7.1. Cut-off frequencies for $LP_{\nu m}$ modes.

$LP_{\nu m}$	$\nu = 0$	$\nu = 1$	$\nu = 2$
$m = 1$	0	2.4048	3.8317
$m = 2$	3.8317	5.5201	7.0156
$m = 3$	7.0156	8.6537	10.1745

References

[1] Marcatili E A J 1969 Dielectric rectangular waveguide and directional coupler for integrated optics *Bell Syst. Tech. J.* **48** 2071–102

[2] Marcuse D 1991 American Telephone and Telegraph Company 1991 *Theory of Dielectric Optical Waveguides (Quantum Electronics Series)* (New York: Academic)

[3] Okamoto K 2010 *Fundamentals of Optical Waveguides* (Amsterdam: Elsevier)

[4] Pollock C R 2003 *Fundamentals of Optoelectronics* (CBLS)

[5] Westerveld W J *et al* 2012 Extension of Marcatili's analytical approach for rectangular silicon optical waveguides *J. Light. Technol.* **30** 2388–401

Part II

Examples of optical sensors with lab exercises

IOP Publishing

Optical Sensors
An introduction with lab demonstrations
Victor Argueta-Diaz

Chapter 8

Laser alignment

The objective of this lab is to learn the basic skills that are often required when working in optical experiments. You will use a beam laser and prepare it for use. Much of this material can be understood through basic geometry (e.g. the manipulations of laser beams as perfectly straight rays). Be careful when handling optical components, not only are they fragile and can be easily damaged, but the lasers that you will be using can damage your eye.

8.1 Justification

Before we work with a laser we need to align it, and collimate it. The reasons for this procedure are:

1. The laser beam is not precisely located (x, y) or pointed (θ_x, θ_y).
2. The beam has unwanted amplitude and phase variations on its profile.
3. The beam diameter is too small.

You will address each of these in turn by:
1. Periscope alignment.
2. Low-pass spatial filtering.
3. Collimation.

8.2 Equipment

- Pinholes: 10, 15 and 25 μm diameters.
- Microscope objectives: 5X, or 10X.
- Collimation lenses: focal lengths of 50, 100 and 250 mm.
- Helium neon laser. Assume a beam spot size of 0.65 mm.
- Spatial filter mount.
- lens mounts.
- Irises.
- Shear-plate collimation tester.
- Detector.

doi:10.1088/978-0-7503-4876-8ch8

8.3 Safety considerations

1. Every time you introduce an object through a laser beam think: where will the reflection go?
2. Use beam blocks to keep the laser beam confined to your optical table.
3. Remove any metallic object that you are wearing (rings, bracelets, necklaces).
4. **NEVER** put your eyes at table level. If you drop a pencil, close your eyes until you are under the table label (do the same on your way up).
5. Never touch a lens (or any optical element) with your bare hands. Some of them have special coatings that can be damaged by the skin oil. Grab all lenses/mirror from the edges. You could use latex gloves, or if you are not confident enough to handle the equipment ask for help.
6. No food or drink in the lab. **Ever**.
7. If you have a cold, bring tissues and use them. Step outside the lab if you need to. Don't coat either the optics or your lab mates with DNA samples. Sneezes and their by-products are the most common contaminant of optical surfaces.
8. Pick up mounted components by the mount.
9. Don't lay optics down on their faces except on lens tissue or in their packages. NEVER lay any optic face down on the steel table—this is begging for scratches and dust.
10. Bolt all mounted components to the table so they can't be knocked over. Designate an out-of-the-way part of the table as 'boneyard' where you keep unused mounted components lightly bolted to the table.
11. The mechanics are moderately hard to damage, but be careful with steel screws—don't over-torque them, don't cross thread them, use washers under screws so the head doesn't chew up the mount.
12. NEATNESS is important. Keep the optical table organized—put all the unused optics in a 'boneyard', lightly screwed to the table. Put your wrenches back in their holder, even during the lab. Keep anything that isn't optics off of the table—notebooks go on the lab benches. Put backpacks and coats somewhere out of the way so you won't trip over them in the dark. Accidents are much more common when a workspace is cluttered. When you are done with the lab, return it to a pristine state—move the optics back to the boneyard, discard any papers or trash and generally tidy up.

8.4 Procedure

8.4.1 Align the laser beam to a desired axis

The first step in any laser experiment is getting the beam to enter your experiment at a known position and angle. Since this includes two transverse positions and two angles, there are a total of four degrees of freedom.

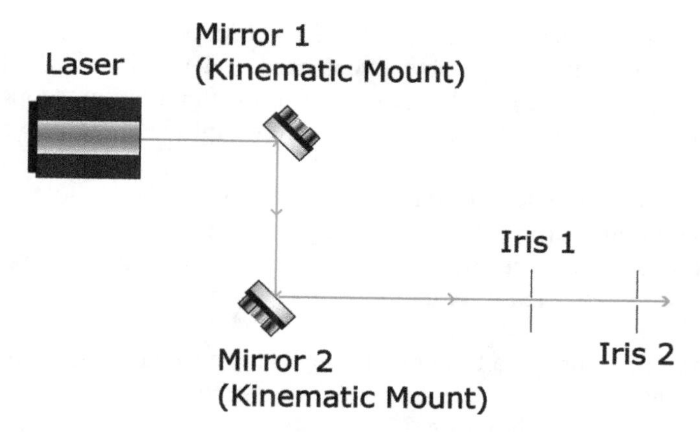

Figure 8.1. Basic periscope configuration. Image designed using ComponentLibrary, created by Alexander Franzen, licensed under a Creative Commons Attribution-Non Commercial 3.0 Unported License. http://www.gwoptics.org/ComponentLibrary/.

Set up two mirrors on tip/tilt mirror mounts in a *periscope* arrangement, as shown in figure 8.1. Mount three identical irises on three identical post holders with no height adjustment so that you know they are exactly the same height. Set one aside for later. Place one iris fairly near M2 and one as far down the table as possible. To get good angular resolution, you'd like as much distance as possible between the two irises. Your job is now to get the laser to go precisely through these two irises.

Adjust the mirrors so that the laser hits the center of the first iris. Close the iris as tightly as possible (but don't break it!) and observe the diffraction rings (a bull's-eye pattern) after the iris using a piece of paper. You want to center this bull's eye on the second iris. You now have four knobs. The system is coupled—each mirror knob adjusts the position on both irises. Iterate until you perfectly center the beam on both irises. **This is the foundation of many experiments, so you want it done well.**

Sketch your layout including distances and components. Briefly describe your alignment procedure for your future reference. Use simple geometry to estimate the position and angular accuracy that you obtained. In the limit of a large distance, the size of the beam on the second iris will increase linearly with distance.

8.4.2 Spatial filtering and collimation

Before we can use the laser beam properly for many of our experiments we need to apply spatial filtering to it. A spatial filter is an optical device that alters the structure of a laser beam. Spatial filtering is commonly used to 'clean up' the output of lasers, removing aberrations in the beam due to imperfect, dirty, or damaged optics, or due to variations in the laser cavity itself.

We will proceed now to introduce a spatial filter into our alignment. Once your laser beam is aligned, you can remove the two irises. Our spatial filter will consist of a microscope objective (usually a 5× or 10× are good choices). You will mount the microscope objective and a pinhole (e.g. 25 μm, on the spatial filter mount. You need to place the pinhole at the focal length of the microscope objective. The location of the focal point of the microscope objective is hard to determine (look at table 8.1 to estimate the focal length and working distances for different microscope objectives), so you will need to adjust the location of your pinhole to improve the spatial filtering.

Use a power meter to detect the maximum output from the pinhole. Once we have obtained a maximum output, we can start collimating our beam.

One of the problems with your current laser beam is that after the pinhole the laser diverges too quickly, and we cannot work with it. Therefore, it is necessary to **collimate** the beam.

A simple way to collimate a laser beam is by using a single aspheric lens, as shown in figure 8.2. The larger the focal length of the aspherical lens, the larger will the beam diameter be after collimation.

Now mount one of the lenses in a lens mount and place the lens a distance equal to its focal length from the pinhole. You can see the expanded beam after the lens using a piece of paper. For a true collimated beam there will be very small changes in its diameter.

Shearing Interferometers can be used to determine if a coherent beam of light is collimated. The design consists of a wedged optical flat mounted at 45° and a diffuser plate with a ruled reference line down the middle. These interferometers are designed to provide qualitative analysis of a beam's collimation.

Table 8.1. Working distance for different microscope objectives.

Power	NA	f (mm)	Working dist. (mm)
5×	0.10	25.4 mm	13.0
10×	0.25	16.5	5.5
20×	0.40	9.0	1.7

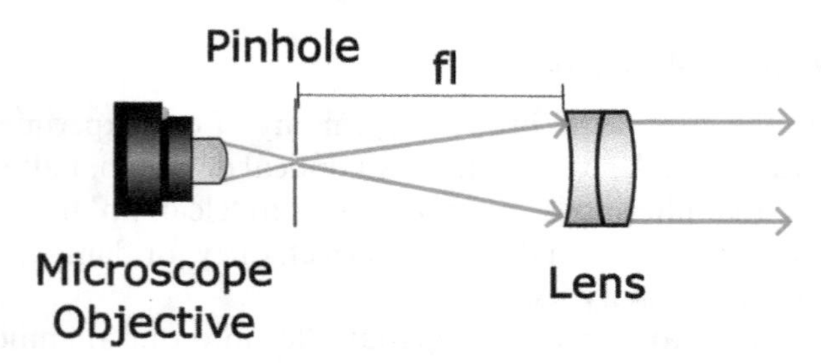

Figure 8.2. Simple laser beam collimation. The distance between the lenses is equal to the sum of the focal length of the two lenses. Image designed using ComponentLibrary, created by Alexander Franzen, licensed under a Creative Commons Attribution-Non Commercial 3.0 Unported License. http://www.gwoptics.org/ComponentLibrary/.

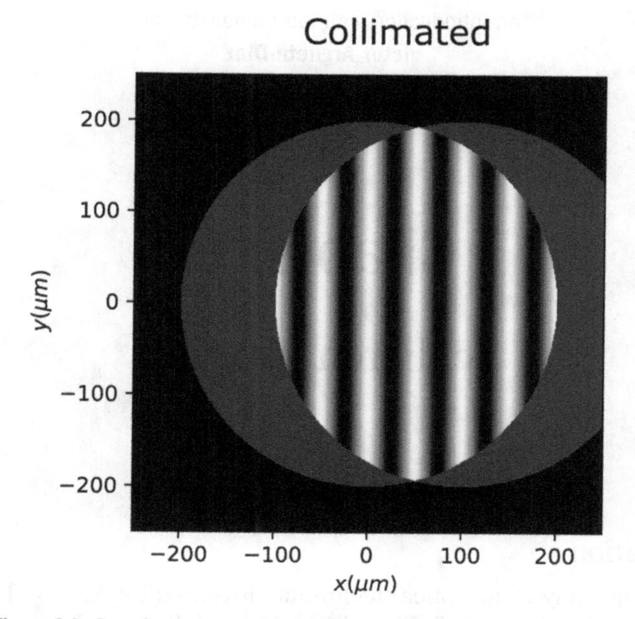

Figure 8.3. Interference pattern on a shear plate for a collimated beam.

The diffuser plate is used to view the interference fringes created by Fresnel reflections from the front and back surfaces of the optical flat. If the beam is collimated, the result microscope pattern will be parallel to the ruled reference line, as shown in figure 8.3. In addition to the degree of collimation, the fringes will also be sensitive to spherical aberration, coma, and astigmatism.

IOP Publishing

Optical Sensors
An introduction with lab demonstrations
Victor Argueta-Diaz

Chapter 9

Schlieren imaging

9.1 Justification

Schlieren photography is an optical technique, invented by August Toepler in 1864, used to display the flow of fluids with different density (i.e. refractive index). Schlieren is an old German word that means 'pieces' in English, however, in optics a schliere is a small region is space which has a different refractive index to its surroundings causing light to refract.

Schlieren imaging is used to visualize density variations in transparent media, such as air or gas. The technique works by using a light source and a lens (or a spherical mirror) to collimate the light from the source. The light passes through the test medium and is refracted due to variations in the refractive index of the medium. The refracted light is then captured by a second lens (or spherical mirror), and focused onto a small point. At this focal point, the beam is partially blocked by a sharp object. Then the light can be observed on a viewing screen or captured by a camera.

The captured image shows variations in the intensity of the light due to the refractive index variations in the test medium, creating a shadowgraph-like image. This allows visualization of otherwise invisible phenomena, such as shock waves, gas flow patterns, and heat convection.

Schlieren imaging is widely used in aerodynamics and fluid mechanics research, as well as in the study of combustion and heat transfer. It can also be used for non-destructive testing of transparent materials, such as glass or plastic.

This technique was use in aeronautical engineering to study the flow or air around objects (think wind tunnels), but it has found application in biology to detect the flow of odors and analyzing breath. Schlieren photography is a simple technique that allows us to detect small changes in refractive index.

doi:10.1088/978-0-7503-4876-8ch9

9.2 Equipment

You will expand the setup that you built during your first lab. In addition you will need:

- a lens;
- razor blade;
- candle, hot gun;
- translation stage.

9.3 Procedure

1. From your first lab, you should have a collimated beam coming from a lens. A collimated beam has a very low divergence, so the beam's spot size should stay the same after a long propagation. Check that you, indeed, have a collimated beam. If you don't, check the position of the lens after the spatial filter, figure 9.1. The distance between the pinhole and the lens should be the lens's focal length.
2. Once the beam is collimated, place a second lens in front of the first one, figure 9.2. The distance between the two lenses shouldn't matter much, but make it at least their combined length (if both lenses have the same focal length, you will have a 4f imaging system). The beam after the second lens will come to focus at the lens's focal point.
3. Place a screen after the image plane, and you should see your spot beam.
4. At the focal point of the second lens, place the razor blade. To do so, mount the blade onto a filter mount. You can use a translation stage such that the blade can completely block the beam spot.
5. Make sure that the translation stage moves perpendicular to the laser beam's propagation direction.

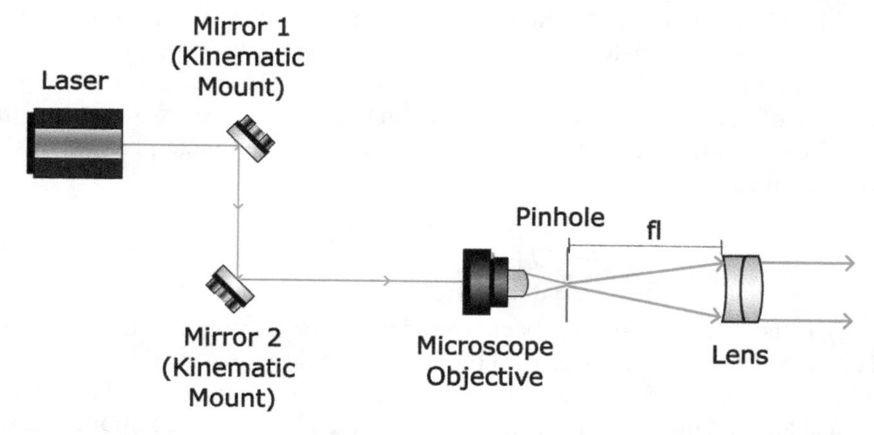

Figure 9.1. Setup after lab 1. Image designed using ComponentLibrary, created by Alexander Franzen, licensed under a Creative Commons Attribution-Non Commercial 3.0 Unported License. http://www.gwoptics.org/ComponentLibrary/.

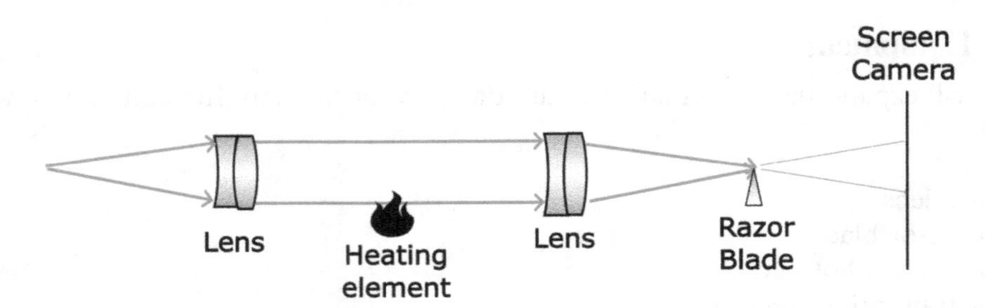

Figure 9.2. Adding a second lens with a knife blade at its focal plane. Image designed using ComponentLibrary, created by Alexander Franzen, licensed under a Creative Commons Attribution-Non Commercial 3.0 Unported License. http://www.gwoptics.org/ComponentLibrary/.

6. Slowly move the blade until it blocks the spot. You should see the shadow of the blade on a screen.
7. Move back your blade until half the beam spot is blocked.
8. You can now include a small heating element between the lenses (e.g., candle, hot gun). On the screen, you should be able to see changes in the shadow of the heating element corresponding to changes in the refractive index of the air.

If you think of light as rays, then the changes in temperature cause the rays to change their path. In some cases, the rays will hit the blade, but in others, they will miss it and be projected onto the screen. Those rapid changes in the ray trajectories are what you see as streaks on the screen.

Schlieren imaging can be used to measure refractive index variations in transparent media, which can provide information about the density variations and gradients in the medium. However, it should be noted that refractive index measurements using Schlieren imaging are typically indirect and require careful calibration of the imaging system.

A calibration technique for Schlieren imaging is called 'background-oriented Schlieren' developed in 1998 by G E A Meier [1], shown in figure 9.3. In this method, a background image is placed between our two lenses. This background image will create a set of reference points that will change positions as the refractive index density of the air in front of it changes. The change in the image position in the y-direction is given by [2]

$$\Delta y = f\left(\frac{Z_D}{Z_D + Z_A - f}\right)\varepsilon_y \tag{9.1}$$

where Z_D is the distance from the background image to the object to be analyzed, Z_A is the distance from the second lens to the object, f is the lens's focal length, and ε_y is the light deflection in the y-direction. A similar expression can be found for the x-direction changes. Once Δy is measured in the image plane, you can approximate the change in refractive index using ε_y and Snell's law.

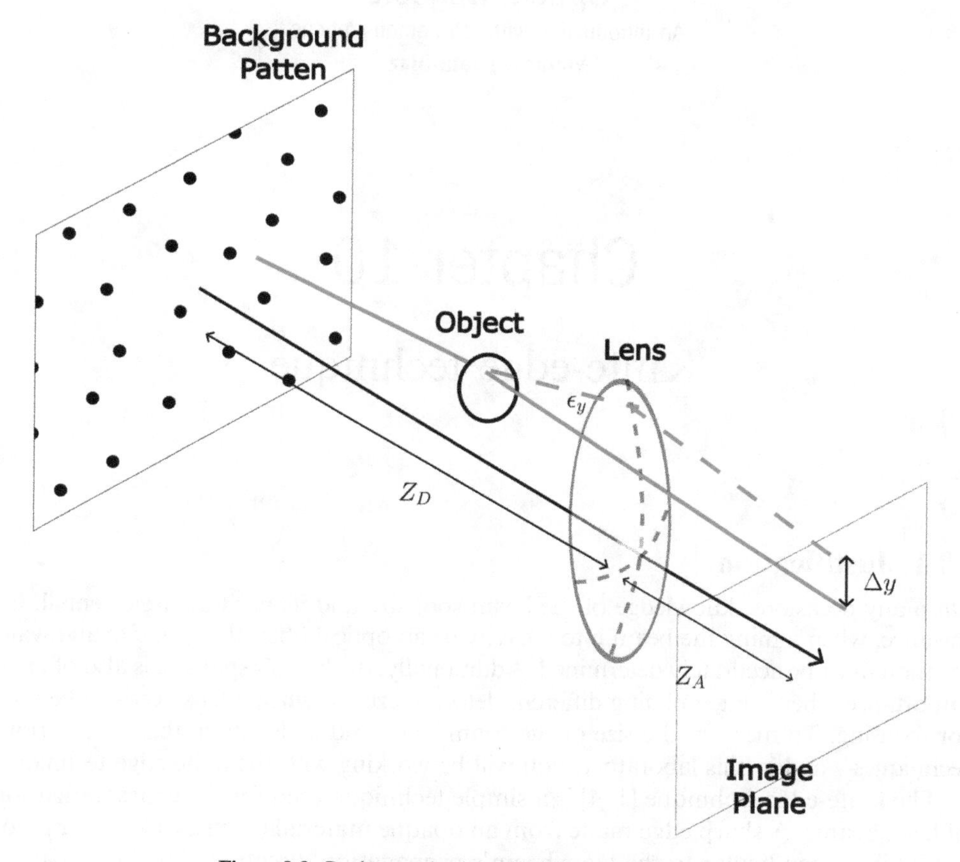

Figure 9.3. Background-oriented Schlieren configuration.

References

[1] Meier G E A 1998 New optical tools for fluid mechanics *Sadhana* **23** 557–67
[2] Raffel M 2015 Background-oriented Schlieren (BOS) techniques *Exp. Fluids* **56/60**

Chapter 10

Knife-edge technique

10.1 Justification

On many occasions, knowledge of the beam spot size and its position are essential; for instance, when aiming the beam into a cavity or an optical fiber, the spot size and waist location must be accurately determined. Additionally, the beam's spot size is also of great importance when contemplating different detector sizes or when a lens needs to be used for focusing. To measure the size of the beam waist and its location, there are various techniques—and in this laboratory, you will be working with the knife-edge technique.

The knife-edge technique [1–4] is a simple technique used for the characterization of laser beams. A sharp edge made from an opaque material (such as a knife or razor blade) is perpendicular to the laser beam's propagation direction. The blade can be moved so that it starts blocking the laser beam, and a detector monitors the transmitted power, a diagram of the technique is shown in figure 10.1. This technique provides a direct way to calculate the spot size, and the location of the beam waist (relative to a lens). Once these parameters are known, the beam waist size (and its position) can be calculated.

10.2 Theory

If we assume a Gaussian beam for the laser beam, then its beam radius changes as the Gaussian beam propagates. The position with the smallest radius is referred to as the waist, denoted by w_0. As distance increases (z), the beam radius, $w(z)$, increases due to divergence, and the Rayleigh range, denoted by z_R, is where the area of the beam is doubled from its area at the waist. This dependence is outlined by the formula:

$$w(z) = w_0\sqrt{1 + \left(\frac{z - z_0}{z_R}\right)^2} \tag{10.1}$$

where z_0 is the location of the beam waist, w_0, z is the distance from the lens to the knife edge, z_R is the Raleigh range.

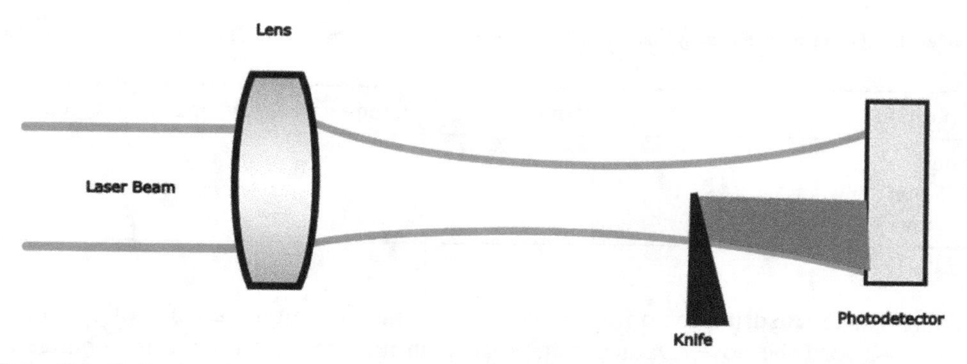

Figure 10.1. Knife-edge technique diagram. Image designed using ComponentLibrary, created by Alexander Franzen, licensed under a Creative Commons Attribution-Non Commercial 3.0 Unported License. http://www.gwoptics.org/ComponentLibrary/.

As the blade is gradually removed from the optical path of the beam, we record the power measured at the photodetector as a function of the lateral position (x) of the blade. The resulting normalized curve can be modeled with the expression shown in equation (10.2).

$$P(x) = \frac{1}{2}\left[1 + \mathrm{erf}\left(\frac{x - x_0}{w(z)}\right)\right] + P_{\mathrm{offset}} \tag{10.2}$$

where:

x is the position of the blade,

x_0 is the location of the beam waist at the z-position,

$w(z)$ is the beam size at the z-position,

P_{offset} is the background power detected by the photodetector in the absence of the laser beam, and erf is the erf function.

In order to obtain the beam parameters, w_0, and z_0, it is necessary to take measurements of the beam size, $w(z)$ at different positions z.

10.3 Equipment

- Oscilloscope;
- Laser (HeNe, 1mW);
- Photodetector;
- Translational stage;
- Knife/Blade.

10.4 Procedure

1. Align the laser beam and power meter so that the power meter measures all of the power emitted by the laser. You can use the procedure of 8 for this step.
2. Focus your collimated laser beam with a 50 mm lens.
3. Mount the razor blade on a translational mount and place it 40 mm from the lens. Make sure that the razor blade is perpendicular to the beam.

Table 10.1. Power after focusing lens for different razor blade positions, x, and different distances from the lens, z.

x-position	40 mm	45 mm	50 mm	55 mm	60 mm
0 mm					
0.25 mm					
0.5 mm					

4. Start measuring the power while moving the blade in increments of 0:25 mm. Record the power values in table 10.1 until there is no longer any change in the measurement power.
5. Move the blade to the next position indicated in table 10.1 and repeat the process.

Once the measurements are completed, we will proceed to calculate the beam spot size at each location of the razor blade (i.e. 40, 45, 50, 55, 60). We will fit the data of the signal detected by the photodetector to equation (10.2). This fitting process will help us obtain the parameters P_{offset}, x_0, and w, at different positions z (we are assuming that we are working with a normalized power).

The fitting process can be done with an excel spreadsheet to evaluate the least-squares values of the three parameters by summing the squared differences between the measured data and the theoretical model. Another method, however, is using a Python code such as the one shown as follows:

```
1  # -*- coding: utf-8 -*-
2  """
3  Created on Fri Jan 13 09:19:33 2023
4
5  @author: arguetav
6  """
7
8
9  import numpy as np
10 import scipy as sp
11 import matplotlib.pyplot as plt
12 from scipy.optimize import curve_fit
13 from scipy import special
14
15
16
17 #example data, alternative it can be parse froma text
       document or similar
18 x_values = np.array([9, 8.75, 8.5, 8.25, 8, 7.9, 7.8, 7.7,
       7.6, 7.5, 7.4, 7.3, 7.2, 7.1, 7.0, 6.9, 6.8, 6.75, 6.7,
       6.6, 6.5, 6.4, 6.3, 6.25, 6.2, 6.1, 6.0, 5.75, 5.5, 5.25,
       5, 4.75, 4.5])
```

Code 10.1. Python code to fit data to Equation 10.2

```
19 beam_data = np.array([2.23, 2.21, 2.17, 2.14, 2.08, 2.04,
       2.00, 1.97, 1.93, 1.89, 1.82, 1.76, 1.66, 1.58, 1.47,
       1.34, 1.21, 1.19, 1.08, 0.95, 0.84, 0.72, 0.61, 0.57,
       0.50, 0.40, 0.33, 0.19, 0.10, 0.10, 0.10, 0.10,0.10])
20
21
22
23
24
25
26 #defining Power function
27 def knife_edge_func(x, x0, w, Poffset):
28     argument = ((x-x0)*np.sqrt(2))/w
29     func = 0.5 * sp.special.erfc(argument) + Poffset
30     return func
31
32
33 #Normalize data
34 beam_data = beam_data/np.max(beam_data)
35
36 ##Calculate middle point of x_values
37 x_values_range = 0.5*(np.max(x_values) - np.min(x_values))
38 x_values = x_values - x_values_range
39
40 #curve fit : function we want to fit, x-values and data
41 #we obtained the w, Pofsset and X0 values as parameters
42
43 fit_params, fit_cov = curve_fit(knife_edge_func, x_values,
       beam_data)
44
45 #generate the fitted curve
46 fit_curve = knife_edge_func(x_values, fit_params[0],
       fit_params[1], fit_params[2])
47
48 #Plot data vs fitted curve
49 plt.figure()
50 plt.grid()
51
52 plt.hlines([0.1, 0.90], x_values.min(), x_values.max(),
       linestyles = 'dashed')
53 plt.plot(x_values, beam_data, '+', label = 'data')
54 plt.plot(x_values, fit_curve, label = 'fitted')
55 plt.legend(loc = 7)
56
57 #Print radius of beam
58 print("Fitted 1/e^2 radius is", np.abs(fit_params[1]), "mm")
```

In figure 10.2 we can see the obtained data and the fitting curved using code 10.1. With this code, we can obtain the $1/e^2$ radius of the beam spot. Another method to obtain the beam spot size is by the x-values where the fitted curved reaches 10% and 90% of the normalized value (shown in the figure as two dashed lines). The

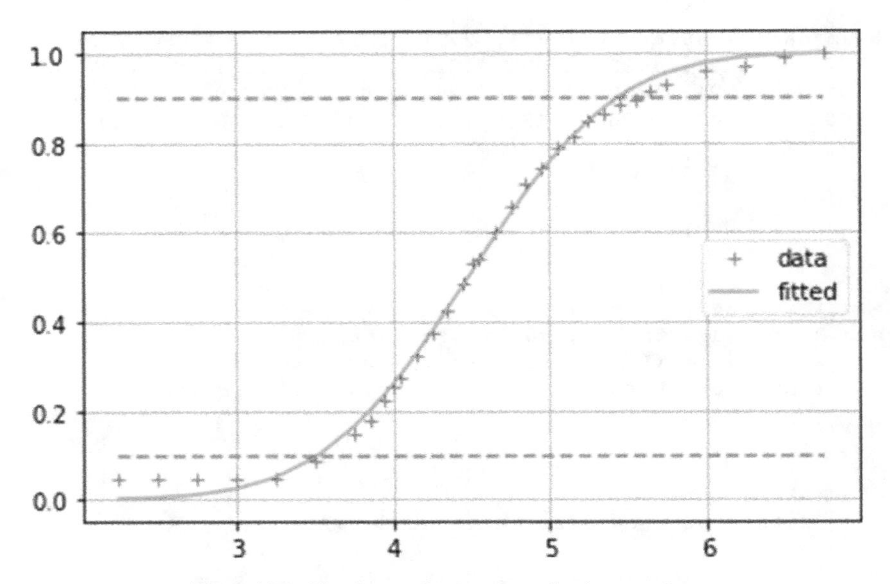

Figure 10.2. Fitted curved using normalized power data.

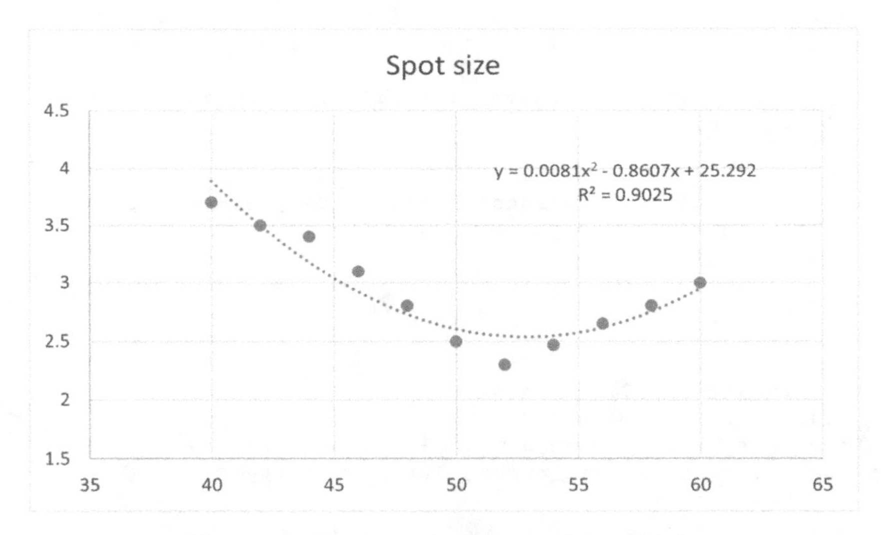

Figure 10.3. Plot of spot size versus z-position of blade.

difference between these two values will give us the beam spot diameter (alternative we could use the 16% and 84% values).

Once significant data points are taken along the z-axis, we can plot the spot sizes as a function of the z-dimension. This will give us an approximation of the spot waist and its location. This can be seen in figure 10.3. In the figure, we did a simple quadratic approximation, but better approximations can be made using the

Gaussian function. We see that the spot size is minimized at a distance of 52 mm from the lens, and its minimum value was around 2.4 mm. This was with a lens of focal length of 50 mm.

10.5 Optical chopper

A faster method of this technique is by using an optical chopper. An optical chopper is a disc that has periodic apertures that interrupt a light beam. As before, the output of the optical detector is displayed on an oscilloscope, and the fall time is a direct measure of the beam size. It is shown by [5] that the time required for the power to drop from 90% to 10% of its full value is related to the Gaussian spot size by:

$$w(z) = 0.7803 \cdot 2\pi f \cdot (t_{10} - t_{90}) \tag{10.3}$$

where:

f is the frequency of oscillation of the chopper,
t_{10} is the time where the power drops to 10% of the total power, and
t_{10} is the time where the power drops to 10% of the total power.

References

[1] de Araújo M A *et al* 2009 Measurement of Gaussian laser beam radius using the knife-edge technique: improvement on data analysis *Appl. Opt.* **48** 393–6

[2] Khosrofian J M and Garetz B 1983 A measurement of a Gaussian laser beam diameter through the direct inversion of knife-edge data *Appl. Opt.* **22** 3406–10

[3] Nemoto S 1986 Determination of waist parameters of a Gaussian beam *Appl. Opt.* **25** 3859–63

[4] Sengupta D and Ung B 2019 Simple optical setup for the undergraduate experimental measurement of the refractive indices and attenuation coefficient of liquid samples and characterization of laser beam profile *Education and Training in Optics and Photonics* 112 (Washington, DC: Optical Society of America) p 11143

[5] Suzaki Y and Tachibana A 1975 Measurement of the um sized radius of Gaussian laser beam using the scanning edge technique *Appl. Opt.* **14** 2809

Chapter 11

Triangulation method

11.1 Justification

Optical techniques for distance measurement offer plenty of uses and applications. Optical techniques are usually the best solution when needing a noncontact method that's fast or automated. In this lab, we explain the triangulation method technique that can be used to measure distances to objects, and related parameters such as displacements, surface profiles, velocities, and vibrations.

We chose this technique because it can be implemented fairly easily in a lab setup at a relatively low cost. However, they are an incredible variety of optical sensing techniques that can be used to obtain the position and velocity of an object. We refer the reader to references [1–4] for further examples and descriptions.

11.2 Theory

The basic principle of a triangulation sensor can be seen in figure 11.1a. A laser source is used to illuminate a spot on the surface of the object we want to measure. A lens is used then to create an image of the spot onto an optical sensor. This sensor can be a CCD/CMOS camera, a photo-array, or any position-sensitive detector (it is even possible to project the spot onto a piece of paper and mark its location). If the surface is displaced so that the light spot is shifted along the path of the laser beam, then the image of the light spot on the detector is also shifted. The displacement of the image on the detector can be used to determine the displacement of the surface. This shift can be calculated using the following equation:

$$\delta = \Delta \cdot M \cdot \sin(\theta) \tag{11.1}$$

where:δ is the position shift at the photodetector,

Δ is the position shift of the surface,

M is the magnification of the system, and

θ is the angle between the beam's optical path and the lens' optical axis.

doi:10.1088/978-0-7503-4876-8ch11

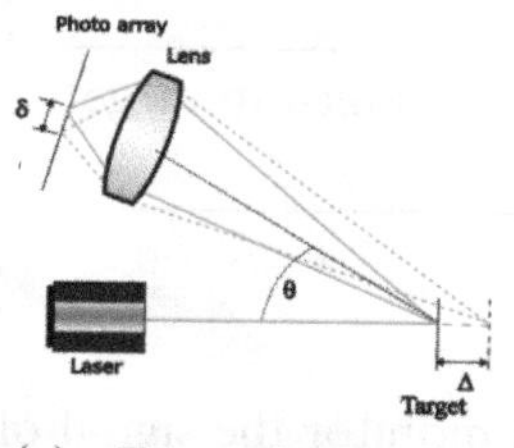

(a) Triangulation setup to measure changes in displacement

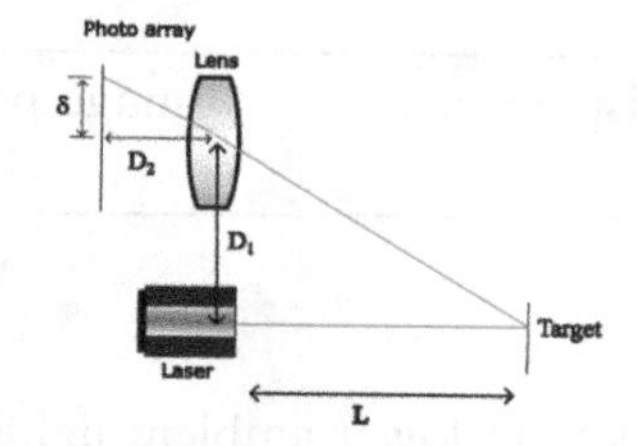

(b) Simply setup to measure absolute distance

Figure 11.1. Simple Triangulation Setup. Image designed using ComponentLibrary, created by Alexander Franzen, licensed under a Creative Commons Attribution-Non Commercial 3.0 Unported License. http://www.gwoptics.org/ComponentLibrary/.

A possible implementation of this position sensor is shown in figure 11.1(b). Here the lens is placed at a known distance D_1 to the side the laser source, and a distance D_2 in front of the photodetector. By measuring the position of the imaged spot on the photodetector, we can calculate the distance to the object by simple geometry:

$$L = \frac{D_1 D_2}{\delta} \tag{11.2}$$

where:
L is the unknown distance to the object,
δ is the position at the photodetector,
D_1 is the lateral distance between the laser source and the lens, and
D_2 is the distance between the lens and the photodetector.

11.3 Equipment

- HeNe laser;
- Concave lens (e.g. 50 mm);
- Photodiode array (e.g. S4111-16Q)[1], or CMOS camera;[2]
- Mounts for equipment;
- One translational mount;
- Oscilloscope;

11.4 Procedure

We will implement the configuration shown in figure 11.1. Place a laser on the optical table and a lens (f = 50 mm), 10 cm to the side of the laser.

Place a target object on a translational mount one-meter in front of the laser. Mount the photodetector behind the lens, and ensure that the laser beam's image on the object is projected onto the photodetector. This image may be very faint, so you

[1] Although a photodiode array has a low cost, it is necessary to implement the necessary electronic circuit to operate this device.

[2] It may be possible to use quadrant sensitive photodiodes, but these tend to be very expensive devices.

Target displacement	Image position	Distance to target

may need to work with lower ambient light, or you can monitor the signal of your photodetector with an oscilloscope or if you are using a USB camera on your computer.

Once you have located the signal on your sensor, translate the target 10 mm forward. Record the new position of the laser beam image. Displace the target four more times in 10 mm increments, recording each image position. Return the target to its initial position and do five more measurements, displacing it five times backward. Fill in the following table, and use equation (11.2) to calculate the distance in the third column

References

[1] Amann M C *et al* 2001 Laser ranging: a critical review of usual techniques for distance measurement *Opt. Eng.* **40** 10–9

[2] Berkovic G and Shafir E 2012 Optical methods for distance and displacement measurements *Adv. Opt. Photonics* **4** 441–71

[3] Girao P M B *et al* 2001 An overview and a contribution to the optical measurement of linear displacement *IEEE Sens. J.* **1** 322–31

[4] Boltryk P J, Hill M, McBride J W and Nascè A 2009 A comparison of precision optical displacement sensors for the 3D measurement of complex surface profiles *Sens. Actuators* A **142** 2–11

IOP Publishing

Optical Sensors
An introduction with lab demonstrations
Victor Argueta-Diaz

Chapter 12

Refractive index and attenuation coefficient

12.1 Justification

The refractive index[1] is a fundamental optical property of many materials. The refractive index of the materials the light passes through can affect different properties like absorption, reflection, transmission, and scattering. It is desired to be able to measure the refractive index of different materials to properly comprehend light–material interactions and create new sensing techniques for their detection.

12.2 Theory

The experimental setup used in this lab for measuring the refractive indices of transparent liquids is based on the experiment shown in figure 12.1 proposed by [1, 2, 4].

In this setup, the displacement (Δ) of a laser beam passing through a cuvette filled with a liquid is related to the value of the its refractive index (n_2) by the following formula:

$$n_2 = n_0 \sin(\theta)\sqrt{1 + \left[\frac{\cos(\theta)}{\sin(\theta) - \frac{\Delta}{d}}\right]^2} \tag{12.1}$$

where:

n_0 is the refractive index of the material surrounding the cuvette (air),
n_2 is the refractive index of the liquid in the cuvette,
θ is the angle between the incident beam the cuvette's face, and
d is the internal dimension of the cuvette.

[1] This lab is a modification from [4]. Reproduced here with the authorization of the authors.

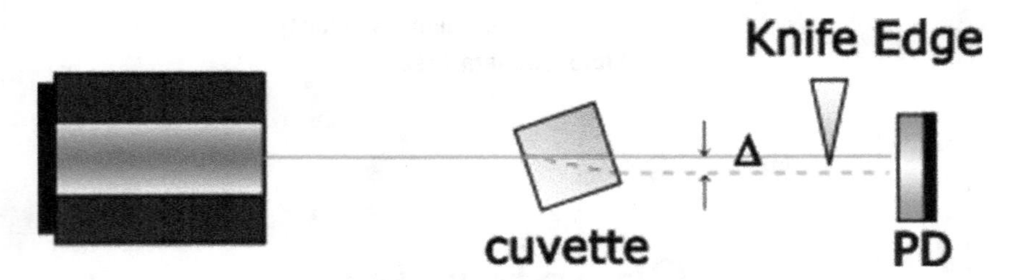

Figure 12.1. Experimental setup for measuring the refractive index of a liquid. Image designed using ComponentLibrary, created by Alexander Franzen, licensed under a Creative Commons Attribution-Non Commercial 3.0 Unported License. http://www.gwoptics.org/ComponentLibrary/.

Therefore, to be able to measure the refractive index n_2 of a semitransparent liquid, we will need to be able to measure the displacement Δ, when a laser beam is incident at angle θ to a cuvette.

12.3 Equipment

1. laser (HeNe 1 mW);
2. Rotating stage;
3. Quartz cuvette;
4. Knife edge;
5. Optical table;
6. Iris (e.g. ID12, Thorlabs);
7. Photo detector.

12.4 Procedure

1. We will first need to characterize the laser beam using the knife edge technique described in chapter 10. We will use the setup shown in figure 12.1, but without the cuvette. The objective is to obtain a profile curve such as the one shown in 10.2.
2. Align an iris with respect of the laser beam and the photo detector.
3. Place an empty cuvette on the rotational stage (align it to the best of your abilities so that laser beam is normal to the cuvette's face).
4. Adjust the angle to the cuvette by sweeping its angular position between ±5° in small increments (e.g., 0.2°). Identify the angle where you record a maximum transmitted power. This will be your angle of reference θ_{ref}. This setup is shown in figure 12.2.
5. Remove the iris and rotate the empty cuvette to an angle of 10°. Characterize the beam using the knife edge technique. Record the transmitted optical power as a function of the blade's position.
6. Replace the empty cuvette with other cuvettes with different liquids, and characterize the beam size for each liquid. Possible options could be distilled water, isopropyl alcohol, or a combination of both at different

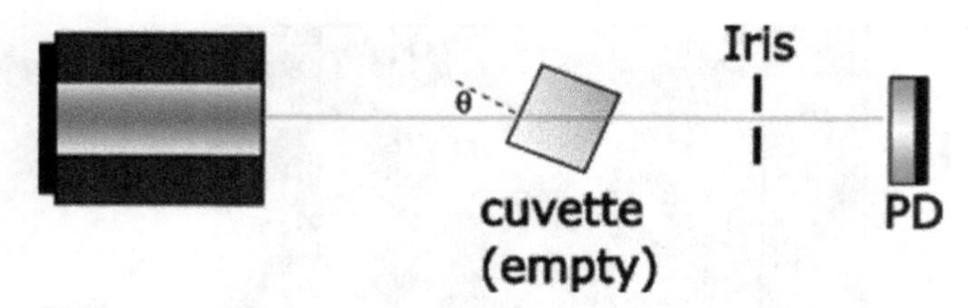

Figure 12.2. Experimental setup to find a cuvette's reference position. Image designed using ComponentLibrary, created by Alexander Franzen, licensed under a Creative Commons Attribution-Non Commercial 3.0 Unported License. http://www.gwoptics.org/ComponentLibrary/.

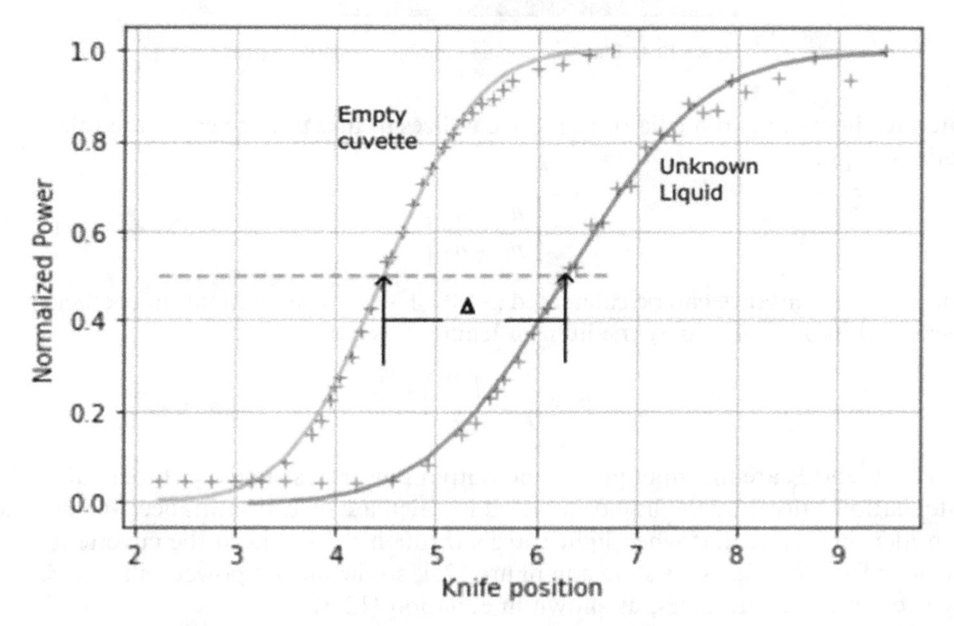

Figure 12.3. Normalized data from an empty cuvette and a cuvette filled with an unknown liquid.

concentrations. Make sure that you don't change the angle of incidence when you replace the cuvette.

7. Normalize each data set to its maximum power and plot each data set in a single graph as a function of the blade's position, as shown in figure 12.3.
8. The beam displacement (Δ) is calculated with respect to the reference curve corresponding to the empty cuvette. We will use the 50% normalized optical power.
9. With the value of Δ for each curve, you can use equation (12.1) to calculate the liquid's refractive index. Be aware that $\theta = 10° - \theta_{\text{ref}}$.

12.5 Attenuation

To measure the attenuation, we will use the transmitted power of the laser beam after it travels through the liquid. The reflectivity at the normal incidence of an

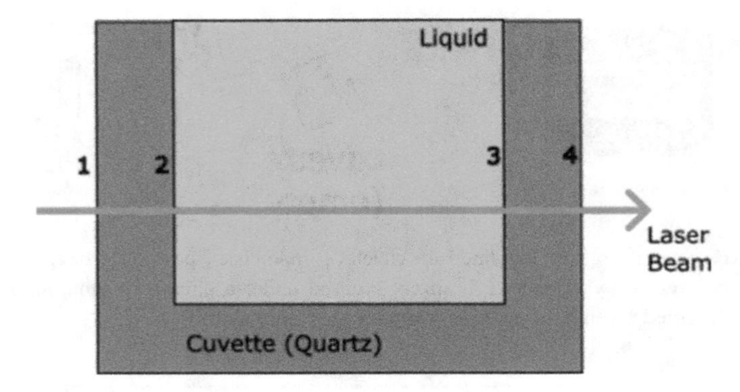

Figure 12.4. Light propagation through a quartz cuvette.

interface between two media of refractive indices n_1 and n_2 is given by the following equation: [3]:

$$R = \left[\frac{n_1 - n_2}{n_1 + n_2}\right]^2 \tag{12.2}$$

and the transmittance can be calculated as $T = 1 - R$. The attenuation coefficient,α, measured in m^{-1}, of a lossy medium of length L is defined by:

$$\alpha = \frac{1}{L}\ln\left(\frac{P_\mathrm{i}}{P_\mathrm{o}}\right) \tag{12.3}$$

where P_i, and P_o are the input power and output power, respectively. To calculate the attenuation caused by the liquid, we need to calculate its transmittance. We need to consider, however, that when light travels through the liquid in the cuvette it goes through four interfaces, as shown in figure 12.4, so the output power will be affected by each of these interfaces, as shown in equation (12.4)

$$P_\mathrm{o} = T_1 T_2 T_3 T_4 T_\mathrm{l} P_\mathrm{i} \tag{12.4}$$

where T_i is the transmittivity at each interface, and T_l is the transmittivity of the liquid.

12.5.1 Procedure

1. Measure the laser power without any cuvette P_i.
2. Place a cuvette with the liquid to measure at the reference angle (normal to the beam path), and measure the optical power, P_o.
3. Using the values of P_i and P_o use equations (12.2) and (12.4) and estimate the value of T_l.

The attenuation of the liquid, will be the attenuation experience by the light traveling from surface 2 to surface 3 in figure 12.4. From equation (12.3) we can say:

$$
\begin{aligned}
\alpha &= \frac{1}{L}\ln\left(\frac{P_\text{i}}{P_\text{o}}\right) \\
&= \frac{1}{L}\ln\left(\frac{T_1T_2P_\text{i}}{T_1T_2T_lP_\text{i}}\right) \\
&= \frac{1}{L}\ln\left(\frac{1}{T_l}\right)
\end{aligned}
\tag{12.5}
$$

References

[1] Nemoto S 1986 Determination of waist parameters of a Gaussian beam *Appl. Opt.* **25** 3859–63

[2] Nemoto S 1992 Measurement of the refractive index of liquid using laser beam displacement *Appl. Opt.* **31** 6690–4

[3] Saleh B E A and Carl Teich M 2019 *Fundamentals of Photonics* (New York: Wiley)

[4] Sengupta D and Ung B 2019 Simple optical setup for the undergraduate experimental measurement of the refractive indices and attenuation coefficient of liquid samples and characterization of laser beam profile *Education and Training in Optics and Photonics* 112 (Washington, DC: Optical Society of America) p 11143

IOP Publishing

Optical Sensors
An introduction with lab demonstrations
Victor Argueta-Diaz

Chapter 13

Polarization and Brewster angle sensor

13.1 Justification

Polarization of light refers to the orientation of the electric field of an electromagnetic wave, such as light, in a specific direction. Light is often described as being unpolarized, meaning that the electric field oscillates in random directions. However, light can be polarized by passing it through materials or by reflecting it off surfaces, causing the electric field to become confined to a single plane.

Polarization sensors are important for a variety of applications, including:

1. **Remote sensing:** polarization sensors can be used in remote sensing applications to determine the orientation and structure of surfaces, such as vegetation, oceans, and atmospheric particles.
2. **Vision systems:** in vision systems, polarization sensors are used to enhance image contrast and reduce glare, making them useful in applications such as machine vision, driver assistance systems, and surveillance cameras.
3. **Robotics:** polarization sensors can be used in robotics to detect and track objects, allowing robots to better navigate and interact with their environments.
4. **Medical imaging:** polarization sensors are used in medical imaging to improve the accuracy and quality of images and to provide new information about biological tissues.
5. **Material** science: polarization sensors can be used to study the optical properties of materials and to investigate the microstructure of materials, such as fibers and crystals.

As mentioned before in chapter 2, there are four different polarizations of light: linear, circular, elliptical and unpolarized. Each has its own distinct electric field vector motion relative to the direction of light propagation. Linear polarization occurs when the electric field vector of light oscillates along a single axis, either horizontally or vertically. Circular polarization occurs when the electric field vector rotates clockwise or counterclockwise in a circular pattern around the direction of propagation. Elliptical polarization occurs when the electric field vector oscillates along two perpendicular axes with unequal magnitudes.

Each type of polarization has unique properties and applications, and the type of polarization can be controlled and manipulated using optical devices such as polarizers, waveplates, and quarter-waveplates.

One consequence of working with different polarizations is that the Fresnel coefficients depend on the polarization of the incident light. When light is polarized, its electric field is confined to a single plane, and this plane of polarization affects the way the light interacts with the boundary between two media.

In general, the Fresnel coefficients for linearly polarized light can be calculated using the refractive indices of the two media and the angle of incidence, and they provide a way to quantify the reflection and transmission of light at a boundary based on its polarization.

The objective of this lab is to use calculate the Fresnel coefficients at the interface with different samples and observe its dependence on light polarization. These sensors are helpful in calculating the refractive index of semitransparent solids, as well as in detecting changes in interfaces between two materials. Furthermore, this setup provides a basic understanding of more complex optical sensors such as ellipsometers and surface plasmon sensors.

13.2 Theory

For linearly polarized light, there are two sets of Fresnel coefficients: the coefficients for *s*-polarized light and the coefficients for *p*-polarized light. *S*-polarized light refers to light with its electric field perpendicular to the plane of incidence, while *p*-polarized light refers to light with its electric field parallel to the plane of incidence [1].

The Fresnel coefficients for *s*-polarized light and *p*-polarized light are different, as the amount of reflection and transmission of light depends on the orientation of the electric field relative to the plane of incidence. For example, *p*-polarized light typically experiences higher transmission and lower reflection than *s*-polarized light when incident on a boundary between two media. These coefficients are:

$$| R_p | = \left| \frac{n_2 \cos(\theta_1) - n_1 \cos(\theta_2)}{n_1 \cos(\theta_2) + n_2 \cos(\theta_1)} \right|^2 \tag{13.1}$$

for the *p*-polarized light, and:

$$| R_s | = \left| \frac{n_1 \cos(\theta_1) - n_2 \cos(\theta_2)}{n_1 \cos(\theta_1) + n_2 \cos(\theta_2)} \right|^2 \tag{13.2}$$

for *s*-polarized light. In these equations, θ_1 and θ_2 are the angles of incidence and refraction, respectively, n_1 is the refractive index of the incidence medium, n_2 is the refractive index of the medium of refraction.

If we substitute Snell's equation, $n_1 \sin(\theta_1) = n_2 \sin(\theta_2)$, into equations (13.1) and (13.2), we can obtain an expression for both coefficients only in terms of θ_1

$$|R_s| = \left| \frac{n_1 \cos\theta_i - n_2\sqrt{1 - \left(\frac{n_1}{n_2}\sin\theta_i\right)^2}}{n_1 \cos\theta_i + n_2\sqrt{1 - \left(\frac{n_1}{n_2}\sin\theta_i\right)^2}} \right|^2, \tag{13.3}$$

and

$$|R_p| = \left| \frac{n_2 \cos\theta_i - n_1\sqrt{1 - \left(\frac{n_1}{n_2}\sin\theta_i\right)^2}}{n_2 \cos\theta_i + n_1\sqrt{1 - \left(\frac{n_1}{n_2}\sin\theta_i\right)^2}} \right|^2 \tag{13.4}$$

The conservation of energy makes it possible to determine the transmitted power simply by calculating the fraction of incident power that does not reflect. So $|T| = 1 - |R|$, for either the *s*- or *p*-polarization coefficient.

In figure 13.1 we can see a graph of the reflection Fresnel coefficients from equations (13.1) and (13.2).

As expected for the *p*-polarization there is angle, the Brewster angle, at which the $|R|$ coefficient goes to zero.

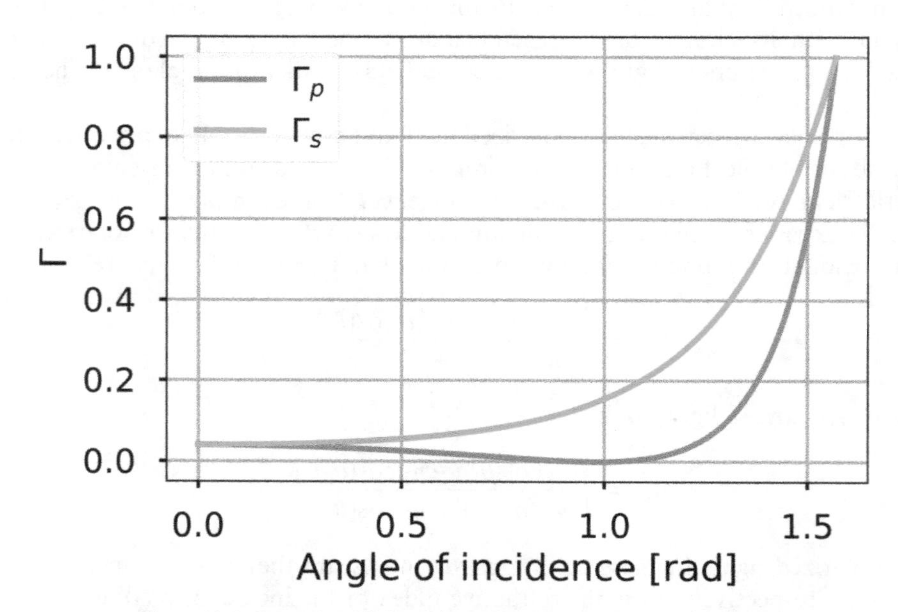

Figure 13.1. Fresnel coefficients (intensity) for a water/glass interface.

We can calculate the Brewster angle, for an air/glass interface, using equation (4.76),

$$\begin{aligned}\theta_{\mathrm{B}} &= \arctan\left(\frac{n_2}{n_1}\right)\\ &= \arctan(1.5)\\ &= 50.31^{\circ}\end{aligned} \tag{13.5}$$

13.3 Equipment

1. Laser (HeNe, 1 mW–5 mW);
2. Linear polarizer (e.g. Thorlabs LPVISE050-A);
3. Rotating stage (e.g. Thorlabs High precision rotational stage PR01);
4. Rotational mount for polarizer (e.g. Thorlabs RSP05);
5. Large area rotating breadboard (e.g. Thorlabs Rotating Breadboard RBB12A/RBB18A);
6. Quartz prism without coatings (at least 25 mm hypotenuse);
7. Prism of different materials;
8. Iris (e.g. ID12, Thorlabs);
9. Photodetector (e.g. PM120D—Digital Power & Console);
10. Miscellaneous mounts, posts and holders as needed.

13.4 Procedure

1. We will work with a collimated beam using the procedure described in the lab experiment in chapter 8.
2. We will use the setup shown in figure 13.2. The collimation setup is not shown here.
3. The photodetector will be mounted on the rotational breadboard, while the prism will be mounted on the rotating stage, allowing for the prism to rotate independently from the photodetector. In the case of Thorlabs's rotating breadboard it is possible to remove the center portion making the setup easier. However, there are different options for performing the same task.
4. Place the linear polarizer between the collimated laser and the prism. Make sure the laser is incident on the prism's hypotenuse, and that you can detect the reflected beam with your power meter. You need to rotate the breadboard, so that you can detect the reflected beam with the photodetector.
5. We will first try to find the Brewster angle. Manually rotate the prism until you identify a minimum power reflected. Note: every time you rotate the prism, you will need to move your photodetector.
6. Rotate the linear polarizer until the light intensity decreases even more.
7. Repeat the last two steps, using the micrometer to rotate the prism until you achieve a minimum light intensity.

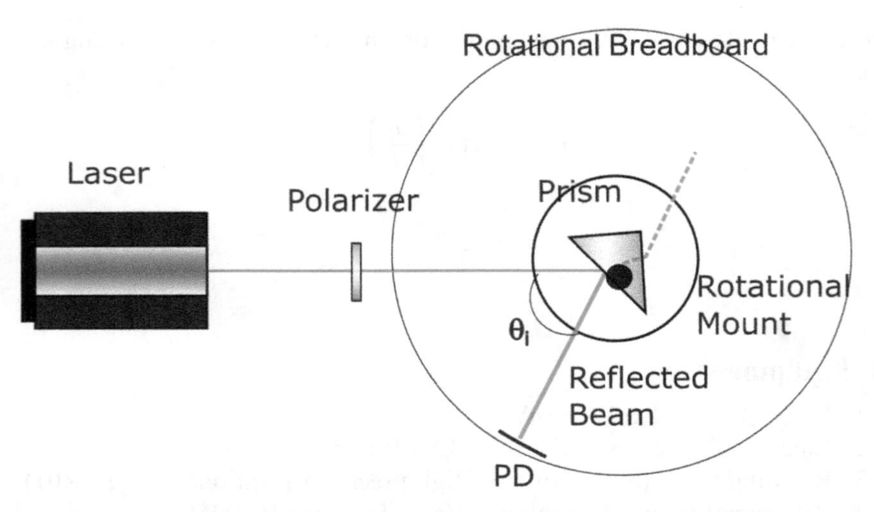

Figure 13.2. Setup for determining Brewster angle. Image designed using ComponentLibrary, created by Alexander Franzen, licensed under a Creative Commons Attribution-Non Commercial 3.0 Unported License. http://www.gwoptics.org/ComponentLibrary/.

8. This will be your Brewster angle, and your *p*-polarization. To find the *s*-polarization, you just need to rotate your linear polarizer 90°.

Reference

[1] Balanis C A 2012 *Advanced Engineering Electromagnetics* 2nd edn (New York: Wiley)

Chapter 14

Michelson interferometer lab

14.1 Justification

The Michelson interferometer uses the principle of interference to measure the properties of light waves. It was invented by Albert Michelson in the late 19th century and has since become a fundamental tool in many areas of physics and engineering.

The Michelson interferometer consists of a light source that emits a beam of coherent light, which is then split into two paths by a beam-splitter. The beam-splitter allows half of the light to pass through, while reflecting the other half at a 90° angle. The two beams of light are then reflected back towards the beam-splitter by two mirrors placed at the ends of each path.

When the two beams recombine at the beam-splitter, they interfere with each other, creating an interference pattern that is observed on a screen placed behind the beam-splitter. The interference pattern is determined by the phase difference between the two beams of light, which is influenced by any changes in the optical path length of one of the beams.

By making precise measurements of the interference pattern, it is possible to determine the phase difference between the two beams and thus measure changes in the optical path length. This makes the Michelson interferometer a powerful tool for measuring small changes in distance, refractive index, and other optical properties of materials.

14.2 Theory

Figure 14.1 depicts a Michelson interferometer in which a laser beam is directed at a beam-splitter. The beam-splitter splits the incident beam into two equal parts, with one part transmitted to a movable mirror (M1) and the other part reflected to a fixed mirror (M2). Both mirrors then reflect the light back towards the beam-splitter. Half of the light from M1 is reflected towards the viewing screen and half of the light from M2 is transmitted through the beam-splitter to the viewing screen.

doi:10.1088/978-0-7503-4876-8ch14

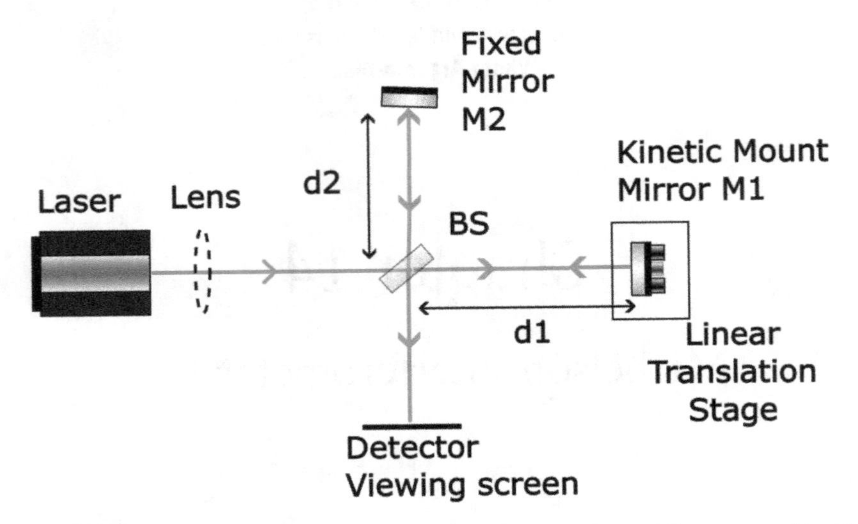

Figure 14.1. Michelson Interferometer setup. Image designed using ComponentLibrary, created by Alexander Franzen, licensed under a Creative Commons Attribution-Non Commercial 3.0 Unported License. http://www.gwoptics.org/ComponentLibrary/.

By adjusting the position of the movable mirror, the path length of one beam can be changed. This results in a modification of the interference pattern, with the locations of maxima and minima switching. By moving the mirror a distance d and counting the number of times the fringe pattern is restored to its original state N, one can calculate the wavelength of the light λ.

$$\lambda = \frac{2\Delta d}{N} \tag{14.1}$$

where Δd is the difference between the final and initial readings of the micrometer, and m is the number of fringe passes.

14.2.1 Measuring refractive index of glass

If we were to introduce a glass plate of thickness t, and refractive index n, parallel to the beam path of mirror M2, there would be an increase of the optical path length of

$$2t(n - 1) = N\lambda \tag{14.2}$$

where N is the number of displaced fringes. If the plate is rotated, the optical path length will increase, and so will the number of displaced fringes. The change in optical path length will depend on the plate's thickness, the rotation angle, and the refractive index. We can figure out the other one if we know two of the three parameters.

If we rotate the glass slide an angle θ, the new optical path length $\overline{AB}$ will be:

$$\overline{AB} = \frac{nt}{\cos(\theta_r)} \tag{14.3}$$

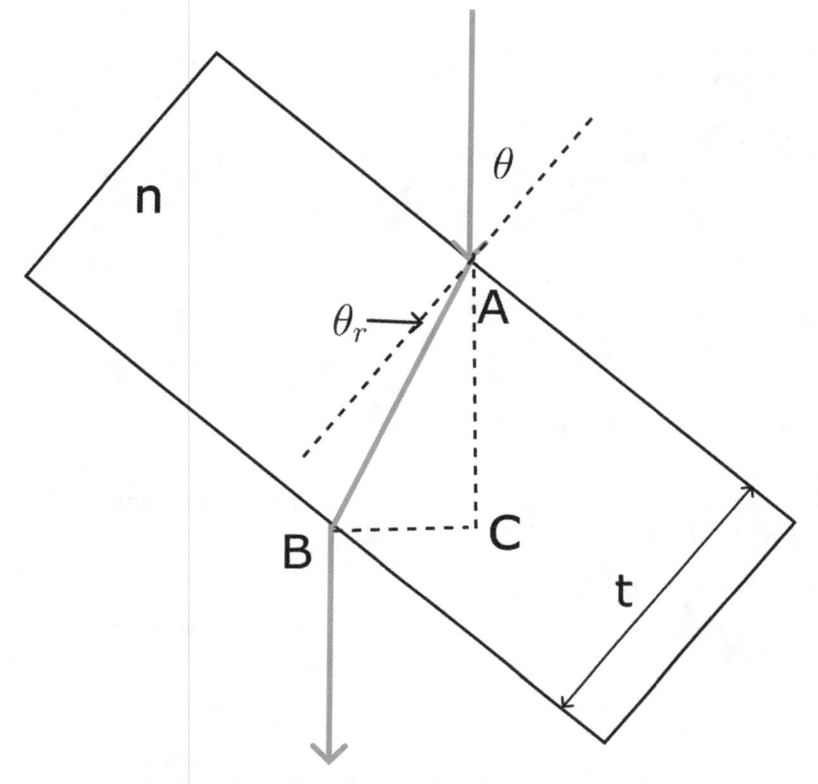

Figure 14.2. Change in optical path length when a glass slide is added.

As shown in figure 14.2. The beam on the other arm, will go through an equivalent path, through air, of just $\overline{AC}$ equivalent to:

$$\overline{AC} = \frac{t\cos(\theta - \theta_r)}{\cos(\theta_r)} \tag{14.4}$$

So the phase difference, δ, caused by the glass slide is:

$$\begin{aligned}\delta(\theta) &= 2k(\overline{AB} - \overline{AC}) \\ &= 2\left(\frac{2\pi}{\lambda}\right)\left(\frac{nt}{\cos(\theta_r)} - \frac{t\cos(\theta - \theta_r)}{\cos(\theta_r)}\right)\end{aligned} \tag{14.5}$$

Notice that if $\theta = 0$, θ_r would also be equal to zero, we will have:

$$\delta(0) = 2\left(\frac{2\pi}{\lambda}\right)(nt - t) \tag{14.6}$$

When we turn the glass plate at an angle θ, we will get a new phase difference of $\delta(\theta) - \delta(0) = 2\pi N$, the number of fringe changes can then be calculated as [1]:

```
1
2 import numpy as np
3 from scipy.optimize import fsolve
4
5 def func(n, N, t, lambd, theta):
6     theta_r = np.arcsin(np.sin(theta) / n)
7     return N - 2 * t / lambd * (n * (1 / np.cos(theta_r) - 1)
      + 1 - np.cos(theta - theta_r) / np.cos(theta_r))
8
9 # Define input parameters
10 t = 1
11 lambd = 546e-6
12 theta = 22.9
13 N=100
14
15 theta_rad = np.radians(theta)
16 # Find solution for n
17 n_guess = 1
18 n_solution = fsolve(func, n_guess, args=(N, t, lambd,
     theta_rad))
19
20 print(n_solution)
```

Code 14.1. Python code to find the refractive index of a slide in a Michelson interferometer.

$$N = \frac{2t}{\lambda}\left[n\left(\frac{1}{\cos(\theta_r)} - 1\right) + 1 - \frac{\cos(\theta - \theta_r)}{\cos(\theta_r)}\right] \quad (14.7)$$

This equation can only be solve numerically. The following code may help you get the right value for the refractive index, we recommend having a large number, $N \approx 50$, for better results.

14.3 Equipment

1. He–Ne Laser;
2. Two flat mirrors;
3. one kinematic mount for a mirror;
4. One fixed mirror mount;
5. A beam-splitter;
6. Precision rotational stage;
7. Linear translation stage;
8. Viewing screen;
9. Lens, 15–25 mm focal length;
10. Microscope slide.

14.4 Procedure

Aligning a Michelson interferometer can be a delicate process and can require some practice and patience. The alignment of the interferometer is critical to ensure that the two beams of light are aligned in such a way that they constructively interfere at the detector.

The key to successful alignment is to make small, incremental adjustments and to use a stable light source, such as a laser. It is also important to use high-quality mirrors and to ensure that the interferometer is free from any external vibrations or disturbances. We will first describe the steps to align the Michelson interferometer

1. Interferometers are very sensitive to vibrations. Try to use a vibration-free surface.
2. Align your laser so that it runs parallel to the optical table. It doesn't need to be collimated or expanded (at this time).
3. We need to place our optical elements as shown in figure 14.1.
4. We will mount one mirror at a time. Place Mirror M1 perpendicular to the laser beam. This is your movable mirror, so make sure that the mirror is mounted on the fixed mount and on top of the linear translation stage.
5. Look at the reflection from the mirror, roughly move the mirror so that the reflected beam goes back to the laser (don't send it back to the laser's cavity).
6. Place your beam-splitter, and make sure that the transmitted beam hits mirror M1 in the same spot as before. Adjust your beam-splitter so that the second beam is at 90° from the first one.
7. Measure the distance between the beam-splitter and mirror M1, that's your path **d1**.
8. Place the second mirror (mounted in a kinematic mount), perpendicular to the second beam. Try to make the distance between the beam-splitter at M2 the same as d1.
9. Mount the viewing screen in front of M2 after the beam-splitter, as shown in the figure.
10. In the viewing screen you should see two spots, move the knobs in mirror M2 until the two spots are overlapping. You may see some flickering on the spot.
11. Add a lens between the laser and the beam-splitter, and adjust its alignment so you can have a larger spot on the viewing screen. You should see some fringes on the spot.
12. Adjust M2 until you get an adequate interference pattern.
13. You can move M1, using the translation stage, and see the effects on the fringe pattern.

14.4.1 Wavelength measurement

In this experiment, we will measure the wavelength of the laser source that has been used so far.

1. Choose a reference point on the interference pattern. It helps if your viewing screen has a millimetric scale, or any kind of marker. If not, you can take a marker to it.
2. Record the position of the micrometer in your movable mirror.
3. Change the distance between the movable mirror and the beam-splitter. While you change the distance, observe your reference point and count until 30 fringes have gone through it.

Table 14.1. Wavelength measurement using the Michelson interferometer.

Measurement	Micrometer initial position	Micrometer final position	Fringe passes	Wavelength
1				
2				
⋮				

Table 14.2. Glass refractive index measurement using the Michelson interferometer.

Measurement	Initial angular position	Final angular position	Fringe passes	Refractive index
1				
2				
⋮				

4. Read the position of your micrometer again.
5. Repeat the last three steps at least five times and record your results in table 14.1.
6. You can calculate the wavelength using equation (14.1).
7. Sometimes, counting the number of fringe passes is hard. If you have a video camera, you can record the interference pattern and later count the passes.

14.4.2 Measuring the refractive index of glass

To measure the refractive index of glass, we need to introduce a glass plate into one of the arms of the interferometer. The glass plate will cause a phase shift of the light passing through it, which will result in a shift of the interference pattern. Here are the steps to measure the refractive index of glass with a Michelson interferometer:

1. We will use a reference point on the viewing screen.
2. Mount the microscope slide on the rotating stage.
3. Place the microscope slide between the beam-splitter and the movable mirror in your interferometer. Have your glass slide perpendicular to your laser beam.
4. Adjust the movable mirror until you get some sharp fringes.
5. Read the angular position on the rotational stage.
6. Slowly rotate your glass slide, and count the fringe passes through your marker. Count at least 100 fringe passes.
7. Read the final angular position of the rotational stage.
8. Repeat the last three steps at least five times. Record your results in table 14.2.
9. To calculate the refractive index of the glass slide use code 14.1.

Reference

[1] Fendley J J 1982 Measurement of refractive index using a Michelson interferometer *Phys. Educ.* **17** 209

Chapter 15

Fabry–Perot interfereometer lab

15.1 Justification

The Fabry–Perot interferometer was invented by French physicists Charles Fabry and Alfred Perot in the late 19th century. The idea was first proposed in 1896 by Fabry, who suggested using multiple reflections between two parallel mirrors to produce interference patterns [1].

Fabry and Perot built their first interferometer in 1899 and used it to study the spectral lines of light emitted by various sources, such as the Sun, stars, and gas lamps. They discovered that the spectral lines were not simple, but instead were made up of many closely spaced lines, which they called 'fine structure'. Their work helped to revolutionize the field of spectroscopy, which is the study of the interaction between light and matter. The Fabry–Perot interferometer provided a much higher resolution than the earlier spectroscopic methods, allowing scientists to observe and measure the fine details of atomic and molecular spectra.

In a Michelson interferometer, for example, a beam-splitter divides an incident light wave into two beams, which are then reflected back by two mirrors to recombine at the beam-splitter. The interference pattern is generated by the combination of these two beams. In contrast, a Fabry–Perot interferometer uses multiple reflections between two parallel mirrors to create an interference pattern. The incoming light is reflected back and forth between the mirrors, creating a standing wave pattern that is dependent on the distance between the mirrors and the wavelength of the light.

The objectives of this lab are:

1. Studying the basic principles of Fabry–Perot interferometry and understanding how the interferometer works to produce interference patterns.
2. Learning how to operate and align a Fabry–Perot interferometer, including adjusting the mirrors and optimizing the input light source.
3. Using the interferometer to measure the wavelength or frequency of light, which can be useful in spectroscopy and telecommunications.

15.2 Theory

A Fabry–Perot interferometer consists of a cavity formed between two highly parallel mirrors, separated by a small distance, as we show in figure 5.11 in our discussion of interference in chapter 5. This cavity allows light to enter and reflect back and forth between the mirrors multiple times before it exits the interferometer. The mirrors are typically coated with a metallic reflective material, such as aluminum or silver, to enhance the reflectivity of the mirrors and reduce the amount of light lost to absorption.

When light enters the cavity, it undergoes multiple reflections between the mirrors, creating multiple beams that interfere with each other. The interference pattern of these beams depends on the wavelength of the light and the distance between the mirrors. Constructive interference occurs when the extra optical path length traveled by a reflected beam is a whole-number multiple of the light's wavelength. Essentially, the multiple reflections in the cavity interfere in the same manner as beams from a multiple-slit grating, with interference maxima becoming increasingly defined as the number of reflections or beams increases.

As we saw in chapter 5, the output intensity of the Fabry–Perot interferometer is given by equation 5.53, which is:

$$\frac{I_t}{I_i} = \mathcal{A} = \frac{1}{1 + [4\mathcal{R}(1 - \mathcal{R})^2]\sin^2(\delta/2)} \tag{15.1}$$

Using the finesse coefficient, we can rewrite the Airy function as:

$$\mathcal{A} = \frac{1}{1 + F\sin^2(\delta/2)} \tag{15.2}$$

where the finesse coefficient is defined as:

$$F = \frac{4\mathcal{R}}{(1 - \mathcal{R})^2} \tag{15.3}$$

It is possible to measure the wavelength of a monochromatic source that illuminates a Fabry–Perot interferometer. Let's assume a cavity with an initial separation between mirror of d_1. If we change the distance between mirrors and count the number of fringes, N, appearing or disappearing at the center we can determine the source wavelength from:

$$\lambda = \frac{2\Delta d}{N} \tag{15.4}$$

15.2.1 Finesse

The finesse of a Fabry–Perot interferometer is a measure of its ability to resolve and distinguish small differences in the wavelength of incident light. It is defined as the ratio of the free spectral range (FSR) to the full-width at half-maximum (FWHM) of the resonant peak of the interferometer.

The free spectral range is the difference in frequency or wavelength between adjacent resonant modes of the interferometer, while the FWHM is the width of the resonant peak at half of its maximum intensity. The finesse is a dimensionless quantity that characterizes the sharpness and selectivity of the interferometer's resonances.

A high finesse interferometer has a narrow linewidth and a high spectral resolution, allowing it to distinguish between closely spaced wavelengths. This property makes it useful for a wide range of applications, such as precision spectroscopy, laser stabilization, and optical communication.

The finesse, $\mathcal{F}$, of a Fabry–Perot interferometer can be calculated using the Finesse Coefficient:

$$\mathcal{F} = \frac{\pi}{2}\sqrt{F} = \frac{\pi\sqrt[4]{R_1R_2}}{1 - \sqrt{R_1R2}} \quad (15.5)$$

where R_1 and R_2 are the reflectivities of the two mirrors.

15.2.2 Free-spectral range

The free spectral range (FSR) in a Fabry–Perot interferometer is the frequency difference between two adjacent resonant modes or wavelengths at which the interferometer transmits or reflects light. It is a fundamental parameter of the interferometer and depends on the distance between the two mirrors and the refractive index of the medium between them.

When light enters the cavity of a Fabry–Perot interferometer, it undergoes multiple reflections between the mirrors, resulting in a series of resonant modes that interfere constructively. Each resonant mode corresponds to a specific wavelength or frequency of the incident light that is resonant within the cavity.

The free spectral range is equal to the difference in frequency or wavelength between two adjacent resonant modes. It can be calculated using the equation:

$$\text{FSR} = \frac{c}{2nd} \quad (15.6)$$

where c is the speed of light, n is the refractive index of the medium between the mirrors, d is the distance between the mirrors, and the factor of 2 accounts for the round-trip distance of light between the mirrors. Notice that FSR is a frequency measurement.

The FSR determines the resolution and sensitivity of the Fabry–Perot interferometer, and it is an important parameter in many applications, including optical filters, wavelength-division multiplexing, laser stabilization, and precision spectroscopy.

We can rewrite our FSR expression in terms of λ, we get [2, 3]:

$$\lambda_{FSR} = \frac{\bar{\lambda}^2}{2d} \tag{15.7}$$

where $\bar{\lambda}$ is the average wavelength of a source with multiple emission wavelengths.

15.3 Equipment

1. Two flat mirrors, protected aluminum coating;
2. He–Ne laser;
3. Sodium light source;
4. Lens, 15–25 mm focal length;
5. Kinematic mount for one mirror;
6. One fixed mirror mount;
7. Linear Translation stage;
8. Viewing screen.

15.4 Procedure

In a sense, it is somehow simpler aligning a Fabry–Perot interferometer than a Michelson or Mach-Zender, if nothing else, just because it has fewer optical elements. However, the alignment of the optical cavity, i.e., the etalon, is critical and requires some patience to achieve a good alignment.

Ideally, the mirrors to be used in a Fabry–Perot interferometer need to have a wedge shape, and high quality surface $\lambda/100$. Unfortunately, these mirrors are usually custom-made and very expensive. It is possible to set up a demonstration Fabry–Perot interferometer with aluminum-protected flat mirrors, that cost less than 30 dollars each. I prefer to use aluminum-protected, over silver or enhanced aluminum, because the interference pattern is easier to observe with a <5 mW laser source. You may need to use a laser with higher output power if using a highly reflective metal.

Another option, is to use a Fabry–Perot interferometer educational kit. They are easier to align and usually include several light sources, but they cost several thousand dollars.

1. Align your laser source so that its beam travels parallel to the optical table.
2. Place one mirror on the kinematic mount, this will be your fixed mirror. The second mirror will be mounted on a non-kinematic mount and on top of the linear translation stage, This will be the movable mirror.
3. Place the fixed mirror ≈50 cm away from the laser, as shown in figure 15.1. Direct the laser beam to the back of the fixed mirror.
4. Place the two flat mirrors one in front of the other, as shown in figure 15.1. Try to have them as parallel as possible. Make the distance between the two mirrors just a couple of millimeters, and be careful not to let the surfaces of the mirrors touch each other.
5. Place the viewing screen as shown in the figure.
6. You should see a set of consecutive beam spots aligned in a straight line, as shown in figure 15.2

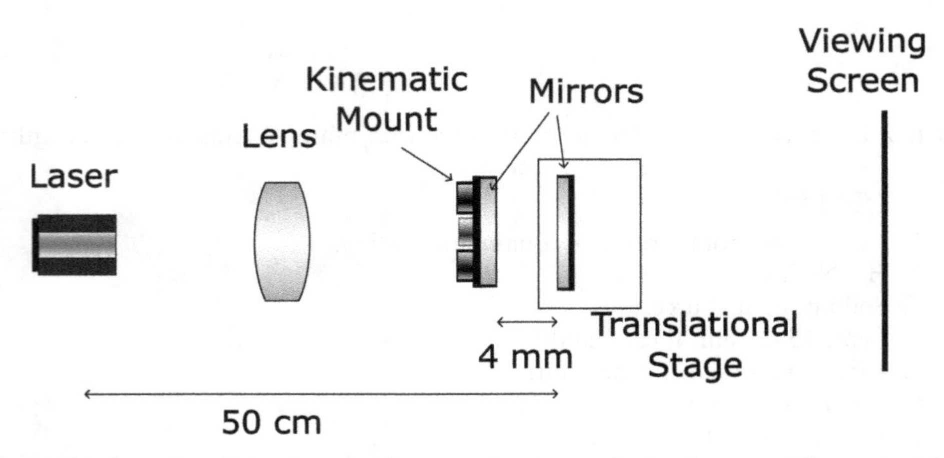

Figure 15.1. Setup for a Fabry–Perot interferometer. Image created using ComponentLibrary, created by Alexander Franzen, licensed under a Creative Commons Attribution-NonCommercial 3.0 Unported License. http://www.gwoptics.org/ComponentLibrary/.

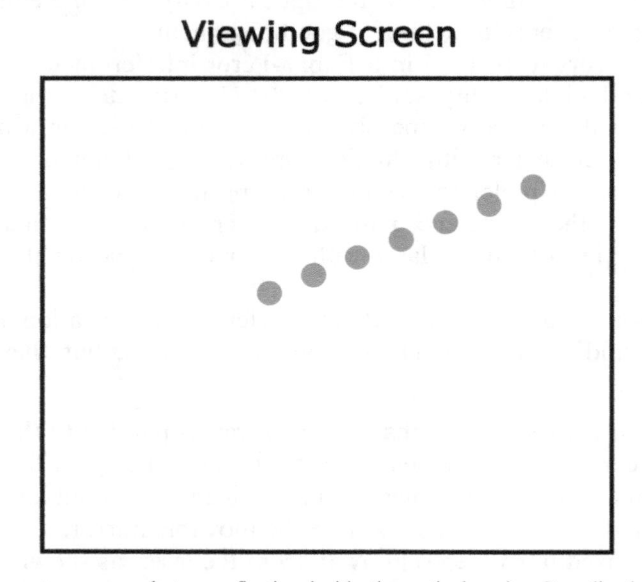

Figure 15.2. Each spot corresponds to a reflection inside the optical cavity. By adjusting the knobs in the kinematic mount, it is possible to overlap all of them into a single spot.

7. By moving the knobs on the kinematic mount, you can change the spacing between the beam spots, and their direction. Adjust the knobs until all spots overlap into a single one. Once this is achieved, both lenses are parallel to each other.

Table 15.1. Table to estimate the wavelength source using the Fabry–Perot interferometer.

Measurement	Fringes (N)	$\Delta d = \lvert d_2 - d_1 \rvert$	λ
1	50		
2	100		
3	150		
⋮			

8. Add a lens between the fixed mirror and the laser source. This lens will diverge the laser beam. Adjust its placement until you see a larger diffraction pattern on the viewing screen. You can also use a 5× microscope objective.

15.4.1 Measurement of source's wavelength

1. Align your Fabry–Perot interferometer so that you observe a well-defined interference pattern.
2. Record the value of the micrometer's position on the movable mirror. This will be your d_1 position.
3. Rotate the micrometer slowly and count the number of fringe changes at the center of the viewing screen. Count at least 50 fringe changes.
4. Record the new position on the micrometer. This will be your d_2 position.
5. Repeat the previous steps, and record your results in table 15.1. Plot the table and find the slope of the graph and use equation (15.4) to determine λ.

15.4.2 Determination of sodium D-lines

A sodium light source consist of two wavelengths that are close to each other, $\lambda_1 = 589.0$ nm and $\lambda_2 = 589.6$ nm. When a sodium lamp is used as the light source, the combined interference fringes of the two spectral lines will appear as two rings close to each other; as you move one of the mirrors, these rings will overlap into a single set of rings. If you continue to change the mirror spacing, the rings will separate again. There is a point where the two sets of rings reach equal separation. If your interferometer doesn't have enough finesse, this will appear as having disappeared or 'wash-out'.

1. Turn the micrometer screw until you find the point of equal separation of the two sodium lines.
2. Record the micrometer reading.
3. Now turn the micrometer until you find the next point of equal separation.
4. Repeat these measurements at least five times. You may have to back the mirror out.
5. Record your results in table 15.2, and use equation (15.7), to calculate the wavelength difference. Assume $\bar{\lambda} = 589.3$ nm.

Table 15.2. Table to calculate the wavelength difference of a sodium source.

Measurement	$\Delta d = \lvert d_2 - d_1 \rvert$	$\Delta\lambda$
1		
2		
3		
⋮		

References

[1] Fabry C and Perot A 1899 Théorie et applications d'une novelle méthode de spectroscopie interférentielles *Ann. Chim. Phys.* **18** 337–78
[2] Hecht J 1991 *City of Light: The Story of Fiber Optics* (Oxford: Oxford University Press)
[3] Pedrotti F L, Pedrotti L M and Pedrotti L S 2017 *Introduction to Optics* (Cambridge: Cambridge University Press)

IOP Publishing

Optical Sensors
An introduction with lab demonstrations
Victor Argueta-Diaz

Chapter 16

Fraunhofer and Fresnel diffraction lab

16.1 Justification

Diffraction is a fundamental physical phenomenon that is important in many areas of physics and engineering. In classical physics, diffraction is described by the Huygens–Fresnel principle that treats each point in a propagating wavefront as a collection of individual spherical wavelets. The characteristic bending pattern is most pronounced when a wave from a coherent source (such as a laser) encounters a slit/aperture comparable in size to its wavelength.

The purpose of this lab is to observe different diffraction patterns created by slits of different widths and shapes.

Diffraction sensors are used in various applications to measure the physical properties of objects or materials, such as their size, shape, and surface roughness. Here are some applications of diffraction sensors:

- **Manufacturing:** diffraction sensors can be used to measure the surface roughness of materials during manufacturing processes, which is important for ensuring product quality and consistency.
- **Medical applications:** diffraction sensors can be used to measure the size and shape of cells and other biological structures, which is important for medical research and diagnostic applications.
- **Material analysis:** diffraction sensors can be used to analyze the crystal structure of materials, which is important for understanding their properties and behavior.
- **Environmental monitoring:** diffraction sensors can be used to measure atmospheric particulate matter, which is important for monitoring air pollution and its effects on human health.

doi:10.1088/978-0-7503-4876-8ch16

Overall, diffraction sensors are important tools for a wide range of scientific and industrial applications, and their use can help us better understand and control the physical world around us.

16.2 Theory

The Fraunhofer diffraction pattern of a single slit is the diffraction pattern that occurs when a collimated beam of light is passed through a single slit and the resulting diffraction pattern is observed at a far distance, where the light can be considered to be effectively parallel. This pattern consists of a central bright maximum, flanked on either side by a series of alternating bright and dark fringes or interference maxima and minima. The central maximum is the brightest, and its width is determined by the width of the slit. The intensity of the other maxima decreases rapidly as the distance from the central maximum increases, and their spacing is determined by the wavelength of the light and the width of the slit.

The equation that describes the diffraction pattern of a single slit was already shown in equation (6.21), we rewrite here, for ease of use.

$$I_{\mathrm{p}} = I_0 \mathrm{sinc}^2\left(\frac{kby}{2z}\right) \tag{16.1}$$

To calculate the size of the central maxima, we need to find the distance between the first two minima, that is: $\Delta y = y_{-1} - y_1$, which gives us:

$$\Delta y = \frac{\lambda z}{b} \tag{16.2}$$

Similarly, the diffraction pattern of a circular aperture in the far-field consists of a bright central spot, surrounded by a series of concentric rings of alternating bright and dark fringes. The size of the central spot is determined by the size of the aperture, while the size and spacing of the rings are determined by the wavelength of the light and the diameter of the aperture.

The central spot of the Fraunhofer diffraction pattern of a circular aperture, called the Airy disk, is much brighter than the surrounding rings and contains most of the energy of the diffracted light. The rings are relatively faint and have a series of dark areas located at regular intervals. The positions of these nulls are determined by the Bessel function. The intensity of the diffraction pattern is calculated by:

$$I_{\mathrm{p}} = 4I_0\left[\frac{J_1(ka\rho/z)}{ka\rho/z}\right]^2 \tag{16.3}$$

As before, we can calculate the diameter of the Airy disk, by calculating when the Bessel function in the previous equation has its first root. This will give us an Airy disk diameter of:

$$\Delta\rho = 1.22\frac{\lambda z}{a} \tag{16.4}$$

Finally, we will try to observe the Fresnel diffraction for a straight edge. This is a phenomenon that occurs when a wave of light passes near a straight edge. Unlike Fraunhofer diffraction, which assumes that the observation point is far away from the diffracting object, Fresnel diffraction takes into account the curvature of the wavefronts of the diffracted light and the distance between the object and the observation point. This makes it a more accurate model for diffraction in close proximity to the diffracting object.

In the case of a straight edge, Fresnel diffraction results in a diffraction pattern that consists of a bright region of light that extends out from the edge, followed by a series of dark and bright fringes that form a series of parallel lines. The spacing of these fringes depends on the distance between the edge and the observation point and the wavelength of the light.

16.3 Equipment

1. Equipment used in Lab 8 for collimation and alignment (optional).
2. Helium-neon laser.
3. Diffraction screens with single-slits of multiple widths[1].
4. Diffraction screens with various circular apertures.
5. Adjustable mechanical slit (optional)[2].
6. 5× microscope objective.
7. Spatial filter system (e.g. Newport Spatial Filter #39-976).
8. Pinhole (25, 50 μm).
9. Optical mounts and holders for the screens.
10. Knife/blade on a mount.
11. Viewing screen.
12. CCD/CMOS Camera + software (optional).

16.4 Procedure

16.4.1 Fraunhofer diffraction single slit

1. One condition that we need to observe Fraunhofer diffraction is that we have plane waves arriving at the diffractive elements. This can be achieved by placing our laser source far away from the diffractive element or collimating our laser using the procedure described in lab 8.
2. In this lab, we will assume that we are placing our source and viewing screen far away from the diffractive element, as shown in figure 16.1. We can satisfy the conditions for Fraunhofer diffraction, for that particular object, only if the distance from both the source and the observing screen is sufficiently large. For a single slit of width b, the Fraunhofer condition without lenses requires both the distances of the source and the observing screen to be much larger than b^2/λ, where λ is the wavelength of our laser source.

[1] There are many vendors, but 3B Scientific offers low-cost alternatives.

[2] Variable mechanic slits offer great versatility, but they may cost several hundreds of dollars, so many times it is not possible to have enough available for a whole class.

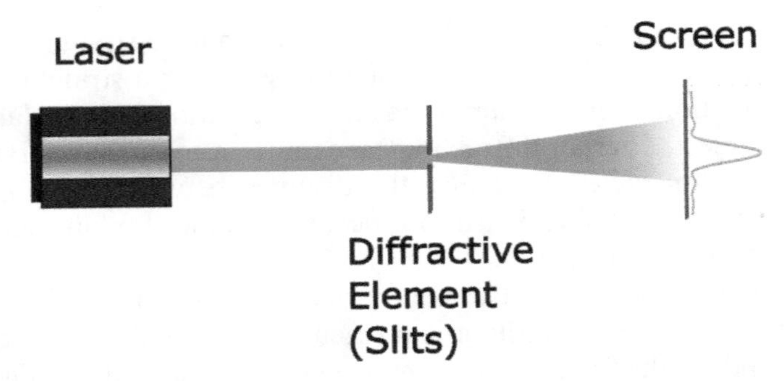

Figure 16.1. Experimental setup for Fraunhofer Experiments. Image designed using ComponentLibrary, created by Alexander Franzen, licensed under a Creative Commons Attribution-NonCommercial 3.0 Unported License. http://www.gwoptics.org/ComponentLibrary/.

Table 16.1. Central maxima dimensions dependence to distance from diffractive element to viewing screen.

Distance (z)	Central maxima dimension (Δy)	Estimated slit width (μm)	Error %

3. Choose one of the diffraction screens with a single slit. Based on the manufacturer's dimensions, estimate the distance that will satisfy the Fraunhofer condition.
4. Place your viewing screen far from the distance you just calculated (a 10× factor should be enough). Use equation (16.2) to calculate the expected separation between the minima on either side of the central maxima in the diffraction pattern. You may need to adjust your distance z so you can observe and measure the central maxima comfortably.
5. We are going to measure the dimension of the central maxima. There are different methods that you can use. A simple way is to display the diffraction pattern on a piece of paper and mark the first minimum location with a pencil.
6. Repeat the last two steps at different distances z. Fill a table of distance versus central maxima dimension (table 16.1).
7. Alternatively, you can use a CCD/CMOS camera to make these measurements. Be aware that many cameras have a protective film over the detector area. This thin film can cause unwanted interference artifacts. You can prevent these artifacts, by projecting the image on a piece of paper or frosted glass and looking at the scattered light.
8. From your measurements, obtain a value for the slit width. Compare your result with the manufacturer's dimensions, and obtain an estimated error percentage.

9. Instead of the slits, mount a piece of hair in a filter holder and place it in front of the laser beam. The pattern is similar to the pattern of a slit. Repeat the steps you just did and calculate the width of the hair.

16.4.2 Fraunhofer diffraction circular aperture

1. You can work with a circular aperture instead of a single slit. The procedure is exactly the same as the one described for the single-slit experiment.
2. We will measure the dimensions of the Airy disk. Use equation (16.4) to estimate this dimension.
3. You can use an iris diaphragm and try to estimate the aperture dimensions based on its diffraction pattern. Although the iris doesn't create a perfectly circular aperture, it can be used as an approximation.

16.4.3 Fresnel diffraction straight edge

To be able to observe Fresnel diffraction, we need to create a point source. We can do this by spatially filtering a laser beam. A laser spatial filter is a device used to improve the quality of a laser beam. It is used to remove spatial distortions or irregularities in the laser beam caused by imperfections in the laser cavity, diffraction, or other sources of interference.

The laser spatial filter works by using a pinhole or a slit to selectively allow only a small, well-defined portion of the laser beam to pass through. This helps to eliminate any spatial variations in the laser beam and produce a more uniform beam with higher quality.

To have a good spatial filter we need to choose the appropriate pinhole diameter for the microscope objective that we are using. The pinhole diameter is a function of the laser wavelength, laser beam diameter, and focal length of the microscope objective used. The diameter can be calculated using the following equation:

$$\text{Pinhole diameter} = \frac{8}{\pi}\frac{\lambda z}{d} \tag{16.5}$$

where z is the focal length of the microscope objective, and d is the diameter of the beam. For a HeNe laser with $\lambda = 635$ nm and a 1 mm diameter beam, we get the following choices for pinhole diameters

1. Use equation (16.5) (or table 16.2) to pick the appropriate pinhole diameter for your laser and microscope objective.

Table 16.2. Pinhole diameter for different microscope objectives.

Objective	Focal length	Pinhole
5×	25.5 mm	50 μm
10×	14.8 mm	25 μm
20×	8.3 mm	15 μm

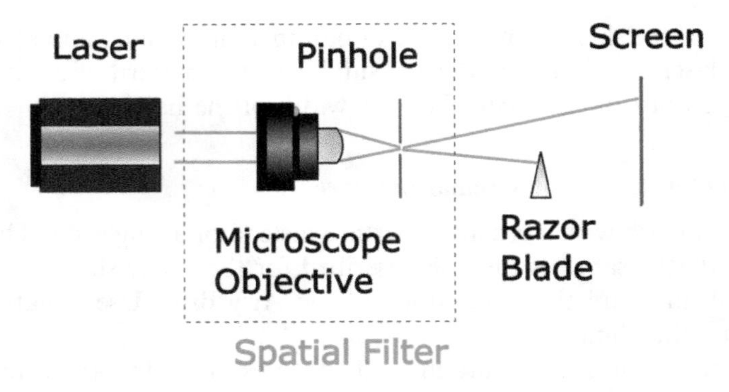

Figure 16.2. Experimental setup for Fresnel Experiments. Image designed using ComponentLibrary, created by Alexander Franzen, licensed under a Creative Commons Attribution-NonCommercial 3.0 Unported License. http://www.gwoptics.org/ComponentLibrary/.

2. Place the pinhole at the focal length of the microscope objective and align the laser beam so that it goes through the center of the pinhole. You can use a photodetector to detect when you get maximum output from your pinhole. Using a spatial filter mount facilitates this alignment considerably.
3. Mount your razor blade on a mount that allows motion in two axes. You want to be able to move the razor blade perpendicular to the trajectory of the laser beam, and parallel.
4. Place the razor blade so that it interrupts the laser beam.
5. Place the viewing screen after the razor blade. Your setup should resemble the one shown in figure 16.2.
6. Move the razor blade until you can observe the Fresnel diffraction pattern in the viewing screen.
7. If you want to observe the Arago Spot, you can replace the razor blade for a small opaque marble. It takes some time to get the right position, but this setup should be adequate to observe this phenomenon.

IOP Publishing

Optical Sensors
An introduction with lab demonstrations
Victor Argueta-Diaz

Chapter 17

Spectrometer lab

17.1 Justification

An optical spectrometer is an optical system instrument used to measure the properties of light over a specific portion of the electromagnetic spectrum. It splits light into its individual wavelengths and measures the intensity of each wavelength. This information can be used to identify the chemical composition of a sample, determine the temperature or pressure of a source of light, or study the properties of a particular material. An optical spectrometer typically consists of several components including:

1. **Light source:** the light source produces the light that is analyzed by the spectrometer. We usually want to use a broadband source.
2. **Entrance slit:** the entrance slit is a narrow opening that allows only a small amount of light to enter the spectrometer.
3. **Collimating lens:** the collimating lens is used to make the light parallel before it enters the diffraction grating.
4. **Diffraction grating**: the prism or diffraction grating is used to separate the different wavelengths of light. The diffraction grating splits the light into its component wavelengths by diffracting the light through a series of parallel grooves.
5. **Focusing lens:** the focusing lens collects and focuses the separated light onto the detector.
6. **Detector:** the detector measures the intensity of the separated wavelengths of light. Common types of detectors used in spectrometers include CCD cameras and photodiodes.
7. **Data acquisition system:** the data acquisition system collects and processes the output from the detector, typically sending the data to a computer for analysis.

doi:10.1088/978-0-7503-4876-8ch17

In this lab we will design and build our own spectrometer. This lab can help you learn about the principles of spectroscopy and how it is used to study the properties of materials. By building a spectrometer, you can observe and analyze the spectral lines of different sources of light, which can help you identify the chemical elements and compounds present in the source.

17.2 Theory

Figure 17.1, shows the basic configuration for a spectrometer. A broadband source is focused onto the slit. This can be done using compound lenses to reduce chromatic aberration, but a single lens with a long focal length, $\approx$100 mm, should also work.

The diverging beam coming from the slit will be collimated by a second lens, and light will be directed to a diffraction grating. The diffraction grating is a critical component of an optical spectrometer. Its primary role is to disperse incoming light into its component wavelengths, allowing the spectrometer to analyze and measure the different wavelengths present in the light source.

Finally a second focusing lens ultimately focuses the light beams onto the detector/screen. The spectrum of the light is obtained by measuring the intensity as a function of position on the detector, with each wavelength occupying a distinct position on the detector.

We will use figure 17.2 to understand the diffraction caused by a diffraction grating. The figure shows a light ray of wavelength λ incident at an angle θ_i and diffracted by a grating of groove spacing d, along a set of angles θ_0, relative to the normal of the grating surface. The diffraction grating equation relates the wavelength of light to the angle of diffraction for a given diffraction grating. The equation is given by:

$$m\lambda = d(\sin\theta_i + \sin\theta_0) \quad (17.1)$$

The resolution of an optical spectrometer refers to its ability to distinguish between two closely spaced spectral lines. The resolution is determined by the spectral bandwidth of the instrument and is typically quantified by the full width at half maximum (FWHM) of the instrument's point spread function.

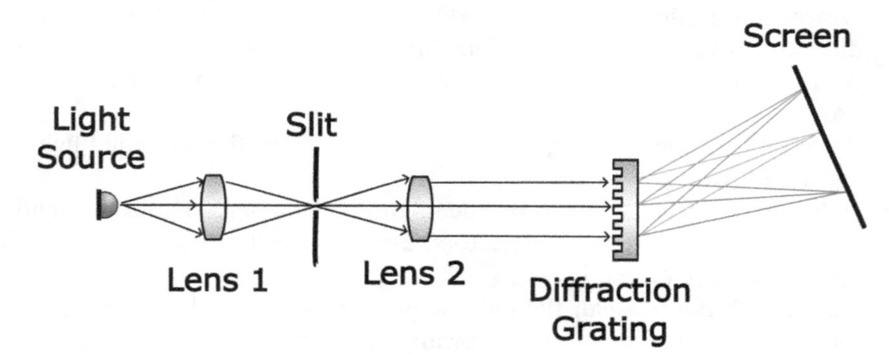

Figure 17.1. Basic Setup for a diffraction grating based spectrometer. Image designed using ComponentLibrary, created by Alexander Franzen, licensed under a Creative Commons Attribution-NonCommercial 3.0 Unported License. http://www.gwoptics.org/ComponentLibrary/.

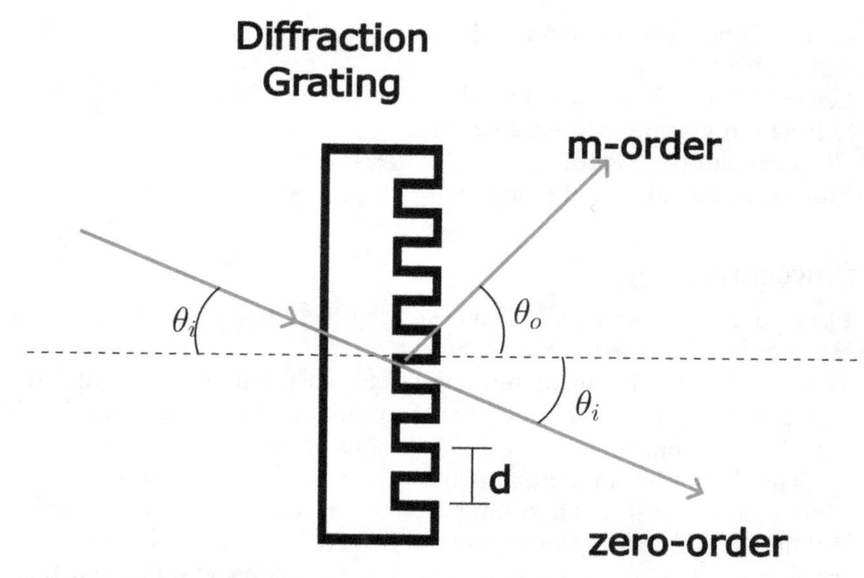

Figure 17.2. Transmission Diffraction Grating. Image designed using ComponentLibrary, created by Alexander Franzen, licensed under a Creative Commons Attribution-NonCommercial 3.0 Unported License. http://www.gwoptics.org/ComponentLibrary/.

To calculate the resolution of an optical spectrometer, you can use the following equation [1]

$$R = \frac{\lambda}{\Delta\lambda} = Nm \tag{17.2}$$

where: R is the resolution of the spectrometer, $\Delta\lambda$ is the spectral bandwidth of the instrument, which is typically defined as the FWHM of the instrument's point spread function, N is the number of grooves on the diffraction grating, and m is the diffraction order of the spectral line being measured.

In order to use the previous equation, we must illuminate the diffraction grating completely. If this is not possible, we can modify the previous equation to:

$$R = \frac{W \sin\theta_m}{\lambda} \tag{17.3}$$

where W is the length of the diffraction grating that is being illuminated, and θ_m is the m-order angle of diffraction.

17.3 Equipment

1. Halogen-Tungsten light source. Economy options like Thorlabs QTH10, should be enough for demonstration purposes.
2. LEDs of different colors.
3. Mercury or sodium lamps, the additional light sources will allow the user to observe different light spectrums.

4. Laser of known wavelength.
5. Adjustable slit.
6. Lenses: 50 mm fl, and 100 mm fl).
7. Diffraction grating, 600 lines per mm.
8. CCD camera (optional).
9. Mounts and holders as needed.

17.4 Procedure

1. Place your source and slit in front of each other, approximately 30 cm apart. Have both of them at the same height.
2. Place a 50 mm lens in the midpoint between the light source and the slit. Adjust the lens position until you see an image of the source on the slit.
3. Place the 100 mm lens behind the slit at approximately 10–12 cm. Light after the lens should be collimated. Adjust the lens position as needed.
4. Illuminate the diffraction grating. Try to cover as much area as possible.
5. Rotate the diffraction grating so that the zero-order beam is straight behind the grating. If you are using a reflective diffraction grating, aim the zero-order beam back to the entry list. The objective is to place the diffraction grating at 0° for the input beam.
6. Adjust the position of the second lens until you get a sharp image of the slit at the zero-order beam.
7. Locate the first-order beams and place the viewing screen.

Once you have your spectrometer aligned, we need to calibrate it. This step involves establishing a relationship between the position on the viewing screen and the corresponding wavelength of light. You need a light source that emits a known wavelength to do this. Once it passes through the spectrometer, you can locate the position of the first-order beam on the viewing screen.

You can now replace your light source with different ones. For example, mercury lamps have emissions at different wavelengths. Use your reference point and equation (17.1) to calculate the emissions wavelengths. Record your results in table 17.1.

Table 17.1. Mercury emission wavelengths.

Wavelength	Calculated	Error
405 nm		
436 nm		
492 nm		
546 nm		
578 nm		
623 nm		
691 nm		

Reference

[1] Pedrotti F L, Pedrotti L M and Pedrotti L S 2017 *Introduction to Optics* (Cambridge: Cambridge University Press)

Part III

Applications of optical sensors

IOP Publishing

Optical Sensors
An introduction with lab demonstrations
Victor Argueta-Diaz

Chapter 18

Light detection and ranging (LiDAR)

18.1 Introduction

LiDAR (Light Detection and Ranging)[1] is a remote sensing technology that uses lasers to measure distances and create 3D representations of the environment. It works by emitting a laser pulse toward an object and then measuring the time it takes for the light to bounce back to the sensor. By repeating this process many times per second, LiDAR can create a dense point cloud of measurements representing objects' shape and location in the surrounding environment. LiDAR is commonly used in applications such as autonomous vehicles, aerial mapping, surveying, and archaeology.

LiDAR technology was first developed in the 1960s, primarily for military applications such as terrain mapping and target detection. The earliest LiDAR systems used large, bulky equipment and were operated from airplanes and helicopters [20].

One of the key breakthroughs in LiDAR technology came in the 1970s with the development of the first solid-state laser. This made LiDAR systems smaller, more efficient, and more reliable, enabling their use in a wider range of applications. In the 1980s, commercial LiDAR systems began to appear on the market, primarily for surveying and mapping applications. These early systems were still relatively expensive and cumbersome, but they represented a major step forward in LiDAR technology. Since 1990, there have been several important advances in LiDAR technology, including:

1. **MEMS-based scanners:** micro-electromechanical systems (MEMS) based scanners offer improved scanning performance and lower cost compared to traditional mechanical scanners [19].

[1] Sometimes the acronym also stands for Laser Imaging, Detection, and Ranging.

doi:10.1088/978-0-7503-4876-8ch18

2. **Photon-counting detectors:** photon-counting detectors offer improved sensitivity and accuracy compared to traditional avalanche photodiodes (APDs), enabling the detection of weaker return signals [15].
3. **Improved signal processing:** advances in signal processing algorithms have enabled better noise filtering, signal-to-noise ratio enhancement, and more accurate distance measurements [11].
4. **Multi-wavelength LiDAR:** multi-wavelength LiDAR systems use multiple laser wavelengths to provide additional information about the target object's properties, such as its chemical composition and physical structure [18]. Multi-wavelength LiDAR technology uses multiple laser wavelengths to create more accurate and detailed 3D maps of the environment. Multi-wavelength LiDAR technology is expected to be used in various applications such as atmospheric research and remote sensing. The development of multi-wavelength LiDAR technology is expected to drive the growth of the LiDAR market in the next 10 years [17].
5. **Flash LiDAR:** flash LiDAR systems use a single, high-powered laser pulse to capture a 3D image of the environment in a single shot, enabling faster data acquisition and real-time mapping [6].
6. **Integration with other sensors:** LiDAR systems are increasingly being integrated with other sensors such as cameras, IMUs, and GPS receivers to provide a more complete picture of the environment [2, 9].

LiDAR technology serves the purpose of capturing accurate and precise 3D information about the physical environment using laser light. This information is then used to create detailed maps, models, and digital representations of the environment, which can be used for a variety of applications, including urban planning, transportation, forestry, mining, archaeology, and environmental monitoring, among others.

The importance of LiDAR technology lies in its ability to provide high-resolution and accurate data that cannot be obtained through traditional surveying methods. This data can be used to improve the efficiency and safety of various industries, as well as to gain a better understanding of the environment and its changes over time. LiDAR data can also be used for scientific research, such as studying climate change, land cover change, and vegetation dynamics.

In addition, LiDAR technology has become increasingly important in the development of autonomous vehicles and drones, as it enables these vehicles to navigate and avoid obstacles more accurately and safely. Overall, LiDAR technology plays a critical role in many fields, from scientific research to industrial applications, and has the potential to transform the way we interact with and understand our environment.

18.2 Basic principles

LiDAR technology utilizes the time-of-flight (TOF) principle to capture depth information by measuring the duration of light traveling from a source to an object

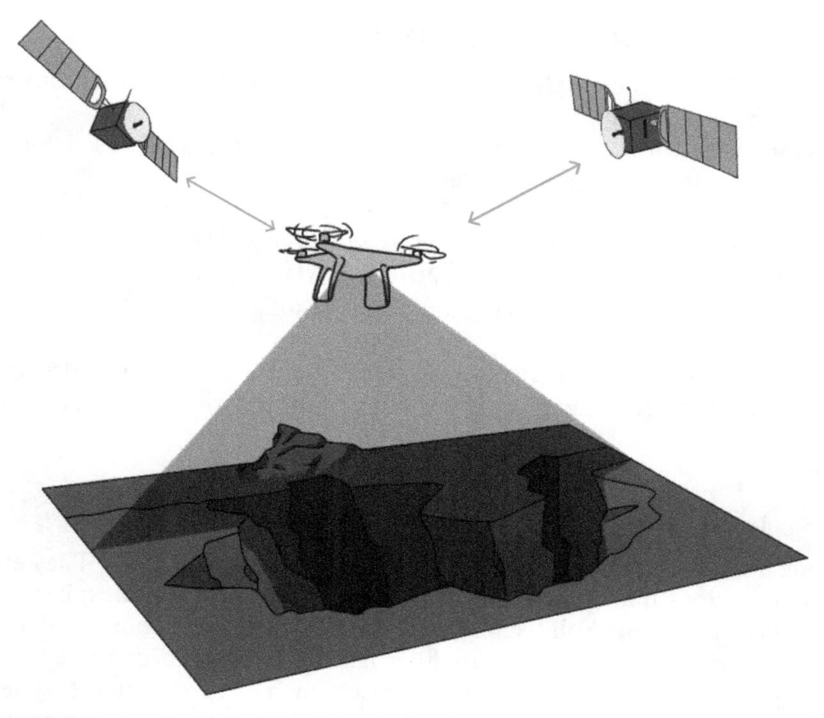

Figure 18.1. Lidar system use for topographic measurements. Image created by L Argueta-Slaughter.

and back. By projecting an optical signal onto an object and detecting the reflected or backscattered signal. The detector measures the time it takes for the laser pulse to travel to the target and back, as well as the intensity of the reflected light, records the laser angle, and from this information, computes where the reflecting object is located in three dimensions. LiDAR can calculate the object's distance and generate a 3D point cloud of the object or environment without any physical contact. Figure 18.1, shows a representation of a LiDAR system using for topographic measurement.

The main components of a LiDAR system typically include a laser, scanner, control and data processing unit, Global Navigation Satellite System (GNSS), and Inertial Measurement Unit (IMU). These components work together to emit and receive laser pulses and measure the time delay, and angle of the reflected light to create a 3D representation of the scanned environment.

18.3 Laser sources

LiDAR systems use a variety of laser wavelengths, depending on the application. They typically use infrared lasers with wavelengths from 0.80 to 1.55 micrometers. This wavelength range is chosen because it is well-suited for atmospheric transmission and is not visible to the human eye.

Other wavelengths that are used in LiDAR include:

1. Visible light: visible light LiDAR systems are used for applications such as 3D imaging and speed detection.
2. Short-wave infrared (SWIR): SWIR LiDAR systems are used for atmospheric sensing and geological surveying applications.
3. Mid-wave infrared (MWIR): MWIR LiDAR systems are used for fire detection and industrial inspection applications.
4. Long-wave infrared (LWIR): LWIR LiDAR systems are used for applications such as thermal imaging and night vision.

The choice of wavelength depends on various factors such as the reflectance and absorbance of the target, background radiation, atmospheric transmission, and eye-safety concerns.

18.3.1 Solid-state lasers

Solid-state lasers are the most powerful type of laser used in LiDARs. They are often used in applications that require long-range detection, such as atmospheric sensing and geological surveying. Solid-state lasers work by pumping a gain medium with a high-energy laser, such as a diode or fiber laser. The gain medium is typically a crystal or glass doped with a rare-earth element, such as neodymium or ytterbium. When the gain medium is pumped, it produces a coherent beam of light that can be used for LiDAR [11, 16].

Solid-state lasers have a number of advantages over other types of lasers, including:

1. High power output: solid-state lasers can produce very high power outputs, which is necessary for LiDAR applications that require long-range detection.
2. Long lifetime: solid-state lasers have a long lifetime, which makes them a good choice for LiDAR applications that require a high level of reliability.
3. Good beam quality: solid-state lasers can produce very good beam quality, which is important for LiDAR applications that require high accuracy.

However, solid-state lasers also have some disadvantages, including:

1. High cost: solid-state lasers are more expensive than other types of lasers, such as diode lasers.
2. Size and weight: solid-state lasers can be large and heavy, which can be a disadvantage for some LiDAR applications.
3. Sensitivity to environmental conditions: solid-state lasers can be sensitive to environmental conditions, such as temperature and humidity.

Despite these disadvantages, solid-state lasers are a good choice for a variety of LiDAR applications that require long-range detection and high accuracy.

18.3.2 Fiber laser

Another common light source used in LiDARs is fiber lasers. Fiber lasers are a type of solid-state laser that uses an optical fiber as the gain medium. This makes them compact, reliable, and efficient. Fiber lasers are also available in a wide range of wavelengths, making them suitable for various lidar applications (figure 18.2).

Fiber lasers are often used in LiDAR applications that require high performance, such as self-driving cars and 3D mapping. They are also used in LiDAR applications that require long-range detection, such as atmospheric sensing and geological surveying [8, 11, 20].

Here are some of the advantages of using fiber lasers as light sources for LiDARs:

1. Compact and reliable: fiber lasers are compact and reliable, which makes them a good choice for LiDAR applications that require a high level of portability and reliability.
2. Efficient: fiber lasers are efficient, which means that they can produce a lot of power from a small amount of input energy. This is important for LiDAR applications that require long-range detection.
3. Available in a wide range of wavelengths: fiber lasers are available in a wide range of wavelengths, making them suitable for various LiDAR applications.

Here are some of the disadvantages of using fiber lasers as light sources for LiDARs:

1. High cost: fiber lasers are more expensive than other types of lasers, such as diode lasers.
2. Sensitive to environmental conditions: fiber lasers can be sensitive to environmental conditions like temperature and humidity.

Despite these disadvantages, fiber lasers are a good choice for a variety of LiDAR applications that require high performance and reliability.

18.3.3 Diode lasers

Diode lasers are the most common type of laser used in lidar systems. They are small, lightweight, and relatively inexpensive. They are also available in a wide range of wavelengths, making them suitable for a variety of LiDAR applications.

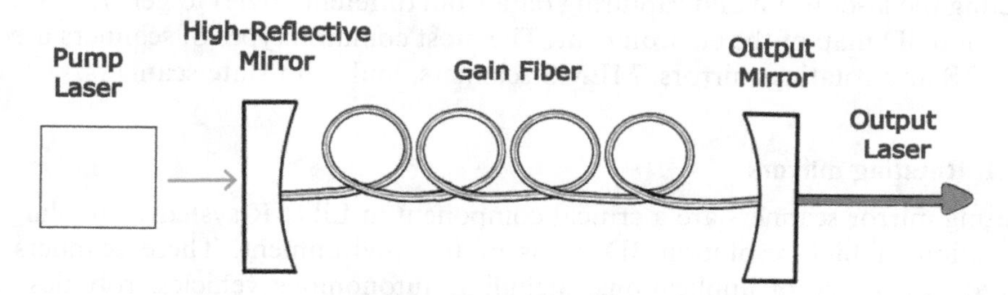

Figure 18.2. Schematic of fiber Laser. Image created by L Argueta-Slaughter.

Diode lasers work by injecting current into a semiconductor diode, which causes the emission of photons. The photons are then collimated and directed into a beam. The wavelength of the emitted photons is determined by the bandgap of the semiconductor material [8, 11, 20].

Diode lasers have a number of advantages over other types of lasers, including:

1. Small size and weight: diode lasers are small and lightweight, which makes them a good choice for LiDAR applications that require portability.
2. Inexpensive: diode lasers are relatively inexpensive, making them a good choice for LiDAR applications with a limited budget.
3. Available in a wide range of wavelengths: diode lasers are available in a wide range of wavelengths, making them suitable for various LiDAR applications.

However, diode lasers also have some disadvantages, including:

1. Low power output: diode lasers typically have a lower power output than other types of lasers such as fiber lasers.
2. Short lifetime: diode lasers have a shorter lifetime than other types of lasers such as fiber lasers.
3. Poor beam quality: diode lasers typically have a poorer beam quality than other types of lasers such as fiber lasers.

Despite these disadvantages, diode lasers are a good choice for a variety of LiDAR applications that require portability and affordability.

18.4 Scanner

The function of the scanner in a LiDAR is to direct and focus the laser beam onto the environment being scanned. A scanner, is used to direct the laser beam at specific angles to cover a particular field of view. As the scanner operates, it directs the laser beam onto the environment, and the scattered beam is captured by a sensor. The scanner's movement allows the LiDAR to capture data from different angles, resulting in a 3D representation of the environment. By measuring the TOF of the laser beam, the distance to objects in the environment can be calculated, enabling the creation of a detailed 3D map of the area. Overall, the scanner plays a critical role in directing the laser beam and capturing data from different angles to generate a high-resolution 3D map of the environment. The most common types of scanners used in LiDARS are: rotating mirrors, MEMS scanners, and solid-state scanners.

18.4.1 Rotating mirrors

Rotating mirror scanners are a critical component in LiDAR systems, enabling the acquisition of high-resolution 3D scans of the environment. These scanners are utilized in a range of applications, including autonomous vehicles, robotics, and environmental monitoring [10, 22] (figure 18.3).

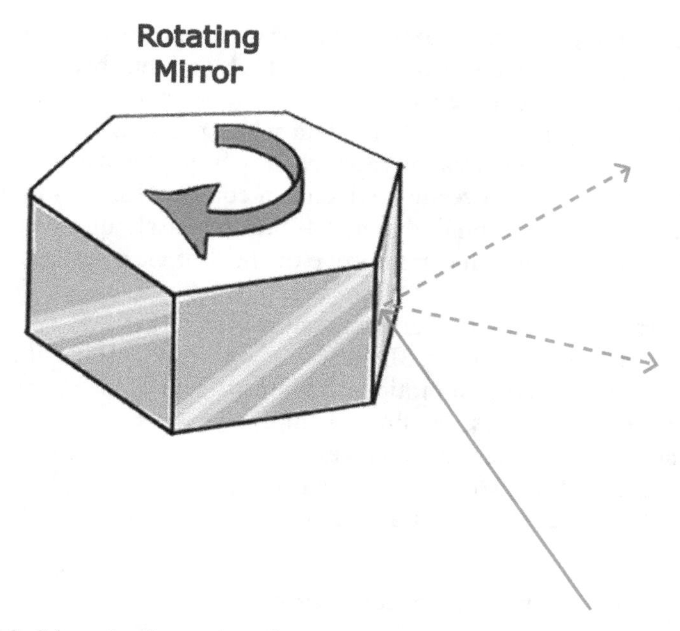

Figure 18.3. Schematic of a rotating mirror scanner. Image created by L Argueta-Slaughter.

Rotating mirror scanners are based on the principle of reflection, where a rotating mirror reflects a laser beam onto an object, and the reflected beam is captured by a sensor. By measuring the TOF of the laser beam, the distance to the object can be calculated, resulting in a 3D representation of the environment. The rotating mirror is typically made of lightweight materials, such as aluminum or glass, and is mounted on a motor that rotates at high speeds, typically between 5 and 20 Hz [19, 21]. The mirror rotates at a constant speed, and the laser beam is directed onto the mirror at a specific angle, resulting in a scan pattern that covers a specific field of view. The captured data is then processed using advanced algorithms to produce a 3D map of the environment.

One of the key advantages of rotating mirror scanners is their ability to generate high-resolution 3D scans in real time. This is particularly important in applications such as autonomous vehicles, where real-time data is critical for safe navigation. Furthermore, the ability to capture data at high frame rates allows for the detection of fast-moving objects, such as pedestrians or other vehicles. Another advantage of rotating mirror scanners is their ability to operate over long ranges, typically up to several hundred meters, making them suitable for a range of outdoor applications, including environmental monitoring and surveying [19, 21].

Despite these advantages, rotating mirror scanners also have some limitations. One of the main limitations is their sensitivity to vibration and motion. Any motion of the scanner, such as vibrations caused by the vehicle or robot, can result in inaccurate data and distortions in the 3D map. Furthermore, the rotating mirror has a limited field of view, typically between 30° and 60°, meaning that multiple scanners may be required to cover a larger field of view. Another limitation is their susceptibility to environmental factors, such as fog or rain, which can affect the accuracy of the data.

Recent research in the field of rotating mirror scanners has focused on improving their accuracy, reliability, and speed. One area of research has been the development of more efficient scanning patterns, such as spiral or random scanning patterns, which can increase the resolution of the data while reducing the time required to scan a given area. Another area of research has been the development of more robust algorithms for data processing, which can compensate for motion and other environmental factors. Additionally, there has been research into the use of multiple rotating mirror scanners, which can improve the field of view and reduce the effects of motion.

Rotating mirror scanners are a critical component in LiDAR systems, enabling the acquisition of high-resolution 3D scans in real-time. While they have some limitations, such as sensitivity to motion and limited field of view, ongoing research is addressing these challenges, leading to improved accuracy and reliability. As LiDAR technology continues to advance, rotating mirror scanners will play an increasingly important role in a range of applications, including autonomous vehicles, robotics, and environmental monitoring.

18.4.2 Micro-electro-mechanical systems mirrors

MEMS (Micro-Electro-Mechanical Systems) mirrors are smaller and less expensive than rotating mirrors, making them ideal for lower-cost LiDAR systems. They consist of tiny mirrors that are controlled by electrostatic or electromagnetic forces. MEMS mirrors can scan the laser beam in different directions at high speeds, but their range of motion is generally limited compared to rotating mirrors [11, 19–21].

MEMS scanners use micro-electromechanical systems technology to move a tiny mirror or lens and direct the laser beam onto the environment being scanned. The mirror or lens can move in two dimensions, typically using electrostatic forces or piezoelectric actuators. By controlling the movement of the mirror or lens, the laser beam can be directed at different angles, covering a particular field of view.

The operation of a MEMS scanner can be explained using the example of a resonant scanner. A resonant scanner consists of a cantilever beam that is attached to a mirror, as shown in figure 18.4. The beam can vibrate at its resonant frequency when an AC voltage is applied to it. The mirror attached to the beam reflects the laser beam onto the environment being scanned, and the reflected light is captured by a sensor. By adjusting the frequency and amplitude of the AC voltage, the resonant frequency and amplitude of the beam can be controlled, allowing the scanner to scan the environment at different frequencies and angles [8, 11].

MEMS scanners offer several advantages over other types of scanners, including:

1. MEMS scanners are much smaller than traditional rotating mirror scanners, making them suitable for miniaturized LiDAR systems. This is particularly important in applications where size and weight are critical, such as unmanned aerial vehicles (UAVs) or wearable devices. MEMS scanners also have a smaller form factor, which allows for more flexibility in the design of LiDAR systems.

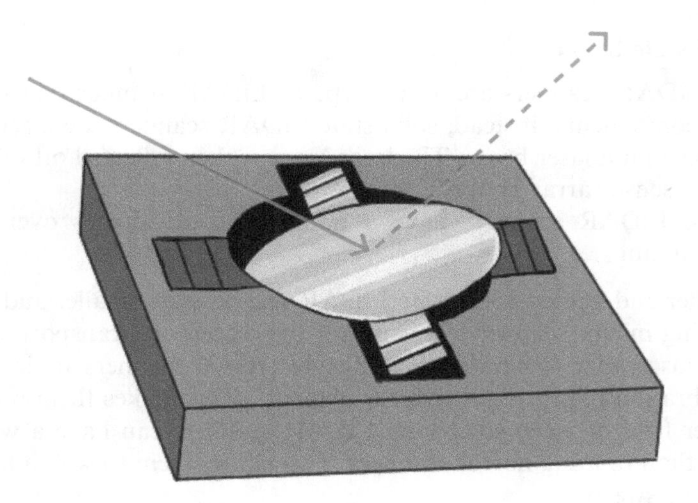

Figure 18.4. Schematic one-dimensional MEMS micromirror. Image created by L Argueta-Slaughter.

2. MEMS scanners require less power to operate than traditional scanners. This is due to their small size and the fact that they use less power to move the mirror or lens. Lower power consumption is particularly important in battery-operated devices, such as UAVs or mobile LiDAR systems, where power consumption is a critical factor.
3. MEMS scanners can respond faster than traditional scanners due to their smaller size and lower mass. This allows them to scan the environment at higher speeds and with greater accuracy, making them suitable for applications where speed and accuracy are critical, such as autonomous driving or robotics [4, 17].

While MEMS scanners offer several advantages over traditional scanners, they also have some limitations, including:

1. MEMS scanners typically have a smaller field of view than traditional scanners. This is due to their small size and limited range of motion. A smaller field of view can limit the ability of a LiDAR system to capture a complete picture of the environment, making them unsuitable for certain applications.
2. MEMS scanners are more expensive than traditional scanners. This is due to the complex manufacturing process involved in producing MEMS devices. The high cost can make them unsuitable for applications where cost is critical.
3. MEMS scanners can be susceptible to environmental factors such as temperature, humidity, and vibration. These factors can affect the accuracy and reliability of the scanner, making it necessary to design the system to minimize their impact [21].

18.4.3 Solid-state lasers

Solid-state LiDAR scanners are a new type of LiDAR scanner that does not use mechanical components. Instead, solid-state LiDAR scanners use a semiconductor laser array to emit a laser beam. The laser beam is then reflected off of objects and detected by a sensor array [1, 14].

Solid-state LiDAR scanners have a number of advantages over mechanical scanners, including:

1. Smaller and lighter: solid-state LiDAR scanners are smaller and lighter than rotating mirror scanners. This makes them easier to transport and deploy.
2. Less susceptible to vibration: solid-state LiDAR scanners are less susceptible to vibration than rotating mirror scanners. This makes them more reliable.
3. Wider field of view: solid-state LiDAR scanners can have a wider field of view than rotating mirror scanners. This allows them to scan a larger area in a single sweep.
4. Higher resolution: solid-state LiDAR scanners can have a higher resolution than rotating mirror scanners. This allows them to create more detailed 3D models of the environment.

As a result of these advantages, solid-state LiDAR scanners are becoming increasingly popular for a variety of applications. For example, solid-state LiDAR scanners are being used in self-driving cars to help them navigate their surroundings. They are also being used in drones to create 3D models of the environment.

Solid-state lasers can achieve very stable, rapid, and precise beam steering. Since there are no mechanical moving parts at all, they are robust and insensitive to external constraints such as acceleration, allowing extremely high scanning speeds (over 100 kHz) over large angles. Moreover, they are highly compact and can be stored in a single chip [15, 18].

Solid-state scanners also have some disadvantages, including:

1. More expensive: solid-state LiDAR scanners are more expensive than rotating mirror scanners.
2. Less mature technology: solid-state LiDAR scanners are a newer technology than rotating mirror scanners. This means that they are less mature and may not be as reliable.
3. Limited availability: solid-state LiDAR scanners are not as widely available as rotating mirror scanners. This can make them more difficult to obtain.

18.4.4 Flash LiDAR

Flash LIDAR technology is a type of LIDAR that operates by emitting a single, powerful laser pulse to illuminate the entire field of view. This single-pulse method is a departure from the traditional scanning LIDAR technology that employs a scanning beam to capture multiple points over time. As a result, flash LIDAR is much faster and more accurate in generating 3D images of the environment [11, 15, 16].

The use of a single pulse also makes flash LIDAR less affected by atmospheric conditions, such as fog or rain, compared to scanning LIDAR systems. This increased robustness allows flash LIDAR to work in a wider range of weather conditions and environments, making it a versatile technology.

The accuracy and speed of flash LIDAR technology make it well-suited for a variety of applications, including autonomous driving, where fast and precise sensing is essential for safe navigation. Additionally, flash LIDAR is used in 3D mapping and industrial inspection applications, where high-resolution imaging and speed are critical. Robotics is another area where flash LIDAR is increasingly being used, particularly in applications where rapid obstacle detection and avoidance are essential. Overall, the many benefits of flash LIDAR are making it an increasingly popular technology in a wide range of industries.

18.5 Other components

18.5.1 Control and data processing unit

Control and data processing units are important components in LiDAR systems that are responsible for controlling the operation of the LiDAR and processing the data that is collected. These units are typically located inside the LiDAR sensor and work together to collect and process LiDAR data in real time [5, 7, 13].

The control unit is responsible for controlling the operation of the LiDAR sensor, including controlling the laser, scanner, and other components. It receives commands from the data processing unit and sends signals to the various components of the LiDAR to control their operation. The control unit also monitors the performance of the LiDAR and provides feedback to the data processing unit to optimize its performance.

The data processing unit is responsible for processing the data collected by the LiDAR sensor. This includes converting the raw data into a usable format, performing data filtering and segmentation, and generating 3D point clouds. The data processing unit also performs post-processing tasks such as data fusion and feature extraction, which are important for many LiDAR applications.

The data processing unit typically consists of a powerful computer that is capable of processing large amounts of data in real time. It is often equipped with specialized software and algorithms that are specifically designed for LiDAR data processing. These algorithms are optimized to handle the unique characteristics of LiDAR data, such as its high spatial resolution and noise characteristics.

One important consideration when designing control and data processing units for LiDAR systems is their power consumption. LiDAR sensors are often used in mobile applications, such as drones and autonomous vehicles, where power is limited. To ensure that the control and data processing units can operate for extended periods of time on limited power, they must be designed to be as energy-efficient as possible.

Another important consideration is the processing speed of the data processing unit. LiDAR sensors can generate large amounts of data very quickly, and the data processing unit must be able to handle this data in real time. This requires a powerful processor and optimized algorithms that can handle the high processing demands of LiDAR data.

18.5.2 Global navigation satellite system (GNSS)

Global Navigation Satellite System (GNSS) refers to a constellation of satellites that provide location and time information to GNSS receivers on the ground. In LiDAR systems, GNSS is used to provide accurate location and timing information for the LiDAR scans [5, 7, 13].

GNSS receivers are commonly integrated into LiDAR systems to provide accurate positioning information. By combining the GNSS position information with the LiDAR point cloud data, it is possible to create highly accurate maps of the target area. This information can be used in a wide range of applications, including autonomous vehicles, robotics, and environmental monitoring.

In autonomous vehicles, GNSS is used to provide location information for the vehicle. This information is combined with the LiDAR data to create a high-precision map of the environment. This map can then be used by the vehicle's onboard computer to navigate and avoid obstacles. GNSS can also be used to provide accurate timing information for LiDAR scans, enabling synchronization with other sensors and improving the accuracy of the LiDAR data.

In robotics, GNSS is used to provide accurate positioning information for the robot. This information can be used to navigate and map the environment, as well as for object recognition and localization. By combining the GNSS data with the LiDAR point cloud data, it is possible to create highly accurate maps of the target area, enabling robots to navigate and operate autonomously in complex environments.

In environmental monitoring applications, GNSS is used to provide accurate positioning information for the LiDAR scans. This information can be used to create accurate maps of the environment, enabling researchers to track changes in the environment over time. GNSS can also be used to provide accurate timing information for LiDAR scans, enabling researchers to synchronize LiDAR data with other environmental data, such as weather and atmospheric conditions.

GNSS plays a critical role in LiDAR systems, providing accurate location and timing information for LiDAR scans. By combining GNSS data with LiDAR point cloud data, it is possible to create highly accurate maps of the target area, enabling a wide range of applications, including autonomous vehicles, robotics, and environmental monitoring.

18.5.3 Inertial measurement unit (IMU)

An inertial measurement unit (IMU) is a device that is commonly used in LiDAR systems to provide accurate positioning and motion data. IMUs consist of accelerometers and gyroscopes, which measure acceleration and rotation, respectively.

Combining the data from these sensors makes it possible to determine the position, orientation, and velocity of the LiDAR scanner. IMUs are often used in LiDAR systems to track the position and orientation of the LiDAR scanner. This information is used to correct for errors in the LiDAR measurements, such as those caused by vibration or changes in the scanner's orientation [5, 7, 13].

IMUs are typically used in conjunction with other sensors, such as GNSS and cameras, to create a complete picture of the environment around the LiDAR scanner. This information can be used for a variety of applications, such as autonomous driving, mapping, and 3D printing. In these systems, the IMU data is combined with data from other sensors, such as GNSS, to provide highly accurate positioning and motion information.

When selecting an IMU for use in a LiDAR system, there are several important characteristics to consider.

The accuracy of the IMU is critical for the overall accuracy of the LiDAR system. It is important to select an IMU that provides high accuracy and stability, with minimal drift over time. This ensures that the motion data collected by the IMU is as accurate as possible, which is essential for producing accurate LiDAR point clouds.

The sensitivity of the IMU is also an important consideration, as it affects the resolution and precision of the motion data. A highly sensitive IMU can detect even small changes in motion, which can improve the accuracy of the LiDAR data. This is especially important for LiDAR systems that require high levels of accuracy, such as those used in aerial mapping and surveying.

The sampling rate of the IMU determines how frequently the motion data is collected. A higher sampling rate can provide more detailed and accurate motion data, but also requires more processing power and storage. This is important to consider when designing the overall LiDAR system, as the sampling rate of the IMU must be balanced against the processing power and storage requirements of the system.

The size and weight of the IMU are also important considerations, especially in mobile LiDAR systems such as those used in autonomous vehicles, drones, and robotics. A smaller and lighter IMU can help to reduce the overall weight and size of the LiDAR system, improving its portability and efficiency. This is particularly important for systems that are designed to be mounted on moving platforms, where size and weight restrictions can be significant.

LiDAR systems can be used in a wide range of environmental conditions, including extreme temperatures, vibrations, and shocks. It is important to select an IMU that is robust and can withstand these conditions without affecting its accuracy or performance. This requires careful consideration of the materials and design of the IMU, as well as its operating specifications and certifications.

The power consumption of the IMU is also an important consideration, especially in mobile LiDAR systems where power is often limited. A low-power IMU can help to reduce the overall power consumption of the LiDAR system, improving its battery life and efficiency. This is particularly important for systems that are designed to operate for extended periods of time without being recharged or refueled.

18.6 Applications

LiDAR technology has various purposes and is important for several applications due to its ability to provide high-precision, three-dimensional data in real time. Here are some of the main applications of LiDAR technology [2, 3, 9, 14, 20, 22]:

1. **Mapping and surveying:** LiDAR technology is used extensively in mapping and surveying applications, particularly for creating digital elevation models and topographical maps. LiDAR data can be used to create accurate, high-resolution 3D models of terrain, buildings, and other objects.
2. **Remote sensing:** LiDAR is used for remote sensing applications to monitor measure various environmental parameters, including forest biomass, land cover, and atmospheric conditions.
3. **Agriculture:** LiDAR technology is also used in agriculture, providing accurate and high-resolution crop monitoring and management data. LiDAR technology is used to create digital terrain models (DTMs) and digital surface models (DSMs), which are used to monitor crop growth and yield. LiDAR technology is also used to create accurate maps of fields, which are used to optimize irrigation and fertilizer application.
4. **Autonomous vehicles:** LiDAR technology is essential for the development of autonomous vehicles. LiDAR sensors are used to provide accurate and high-resolution 3D data, which is used to create a detailed map of the surrounding environment. Autonomous vehicles use this map to navigate and avoid obstacles, making LiDAR technology a critical component of autonomous vehicle technology.
5. **Archaeology and cultural heritage:** LiDAR technology is used in archaeology and cultural heritage applications to document and preserve historic sites and monuments.
6. **Infrastructure and construction:** LiDAR data is used to design and construct infrastructure projects such as bridges, highways, and pipelines. It is also used for monitoring the condition and safety of infrastructure assets over time.
7. **Disaster response:** LiDAR data is used in disaster response efforts to help assess damage and aid in recovery efforts. It can be used to create 3D maps of affected areas and provide data on changes to the landscape after a disaster.

Overall, LiDAR technology has become an essential tool in a variety of fields, from scientific research to commercial and industrial applications. Its ability to provide high-precision, 3D data has revolutionized the way we perceive and interact with the world around us.

18.7 Challenges and future perspectives

LiDAR technology is expected to grow significantly in the next 10 years due to advancements in technology, the increasing demand for high-quality data, and the increasing number of applications for LiDAR technology.

The global LiDAR market was valued at USD 1.7 billion in 2020 and is expected to reach USD 4.9 billion by 2025, growing at a Compound Annual Growth Rate of 23.7% during the forecast period [12]. The market is expected to grow due to the increasing demand for LiDAR technology in the automotive, transportation, and construction industries. The demand for LiDAR technology is also increasing due to the need for accurate and high-resolution data for various applications such as mapping, surveying, and urban planning.

Here are some of the factors that are expected to drive the growth of the LiDAR market in the coming years:

1. Increasing demand for 3D data: 3D data is becoming increasingly important for a variety of applications, such as autonomous driving, augmented reality, and virtual reality. LiDAR is a key technology for collecting 3D data, as it can accurately measure the distance to objects in all weather conditions.
2. Development of self-driving cars: self-driving cars need to be able to create a 3D map of their surroundings in order to navigate safely. LiDAR is a key technology for this, as it can accurately measure the distance to objects in all weather conditions.
3. Need for more accurate and detailed mapping: there is a growing need for more accurate and detailed mapping of the Earth's surface. LiDAR is a valuable tool for this, as it can create 3D models of the Earth's surface. This data can be used for a variety of purposes, such as urban planning, disaster management, and environmental monitoring [4].

One of the main challenges of LiDAR technology is cost. While the technology has become more accessible in recent years, high-end systems with high accuracy and resolution can still be quite expensive. This can limit its adoption in certain industries that have tighter budgets.

Another challenge of LiDAR technology is its size and weight. LiDAR systems can be bulky and heavy, which makes them difficult to use in certain applications, such as drones or small robots. This can limit their effectiveness in these areas, where mobility and size are crucial.

LiDAR also has a limited range, particularly in poor weather conditions or low light environments. This can make it difficult to capture accurate data in these situations, which can be a significant limitation in many applications.

Furthermore, LiDAR can generate vast amounts of data, which can be challenging to process and analyze, particularly in real-time applications. This can limit its effectiveness in areas where quick decision-making is crucial.

LiDAR can also be affected by interference from other nearby LiDAR systems or from other sources of electromagnetic radiation. This can reduce its accuracy and make it more challenging to use in certain environments.

While LiDAR can generate highly accurate data, its resolution can be limited, particularly in large-scale mapping applications. This can make it challenging to capture fine details, which can be crucial in certain industries.

Finally, LiDAR can be affected by environmental challenges such as dust, smoke, or fog, which can reduce its accuracy. This can limit its effectiveness in outdoor environments or in areas with high levels of pollution.

Overall, while LiDAR technology has made significant advancements in recent years, there are still challenges that need to be addressed to make it even more widely applicable and effective in various industries.

References

[1] Bhattacharya P 1997 *Semiconductor Optoelectronic Devices* 2nd edn (Englewood Cliffs, NJ: Prentice-Hall)

[2] Bresson G *et al* 2017 Simultaneous localization and mapping: a survey of current trends in autonomous driving *IEEE Trans. Intell. Veh.* **2** 194–220

[3] NOAA Coastal Services Center 2012 *Introduction to Lidar Technology, Data, and Applications* (Charleston, SC: NOAA Coastal Services Center)

[4] Cui Y *et al* 2021 Deep learning for image and point cloud fusion in autonomous driving: a review *IEEE Trans. Intell. Transp. Syst.* **23** 722–39

[5] Dong P and Chen Q 2017 *LiDAR Remote Sensing and Applications (Remote Sensing Applications Series)* (Boca Raton, FL: CRC Press)

[6] Gelbart A *et al* 2002 Flash lidar based on multiple-slit streak tube imaging lidar *Laser Radar Technology and Applications VII* 4723 (Bellingham, WA: SPIE) pp 9–18

[7] Gimmestad G G and Roberts D W 2023 *Lidar Engineering: Introduction to Basic Principles* (Cambridge: Cambridge University Press)

[8] Guo Y *et al* 2020 Deep learning for 3D point clouds: a survey *IEEE Trans. Pattern Anal. Mach. Intell.* **43** 4338–64

[9] Himmelsbach M *et al* 2008 LIDAR-based 3D object perception *Proc. 1st Int. Workshop on Cognition for Technical Systems* 1

[10] Kashani A G *et al* 2015 A review of LiDAR radiometric processing: from ad hoc intensity correction to rigorous radiometric calibration *Sensors* **15** 28099–128

[11] Li Y and Ibanez-Guzman J 2020 Lidar for autonomous driving: the principles, challenges, and trends for automotive lidar and perception systems *IEEE Signal Process. Mag.* **37** 50–61

[12] Markets and Markets 2023 *Which are the major companies in the machine vision market? What are their major strategies to strengthen their market presence?* https://www.marketsandmarkets.com/Market-Reports/lidar-market-1261.html

[13] McManamon P F 2019 *LiDAR Technologies and Systems (Press Monographs)* (Bellingham, WA: SPIE Press)

[14] Mulla D J 2013 Twenty five years of remote sensing in precision agriculture: key advances and remaining knowledge gaps *Biosyst. Eng.* **114** 358–71 (Special Issue: Sensing Technologies for Sustainable Agriculture)

[15] Popescu S *et al* 2018 Photon counting LiDAR: an adaptive ground and canopy height retrieval algorithm for ICESat-2 data *Remote Sens. Environ.* **208** 154–70

[16] Reutebuch S E, Andersen H-E and McGaughey R J 2005 Light detection and ranging (LIDAR): an emerging tool for multiple resource inventory *J. Forest.* **103** 286–92

[17] Roriz R, Cabral J and Gomes T 2022 Automotive LiDAR technology: a survey *IEEE Trans. Intell. Transp. Syst.* **23** 6282–97

[18] Wandinger U *et al* 2002 Optical and microphysical characterization of biomass-burning and industrial-pollution aerosols from-multiwavelength lidar and aircraft measurements *J. Geophys. Res. Atmos.* **107** LAC–7

[19] Wang D, Watkins C and Xie H 2020 MEMS mirrors for LiDAR: a review *Micromachines* **11** 456

[20] Wang X *et al* 2020 The evolution of LiDAR and its application in high precision measurement *IOP Conf. Ser.: Earth Environ. Sci.* **502** 012008

[21] Woong Yoo H *et al* 2018 MEMS-based lidar for autonomous driving *e & i Elektrotech. Informationstech.* **135** 408–15

[22] Zamanakos G *et al* 2021 A comprehensive survey of LIDAR-based 3D object detection methods with deep learning for autonomous driving *Comput. Graph.* **99** 153–81

IOP Publishing

Optical Sensors
An introduction with lab demonstrations
Victor Argueta-Diaz

Chapter 19

Optical biosensors

19.1 Introduction

Without a doubt, optical sensors have an important presence in the fields of chemistry and biochemistry. Specifically in the creation of biosensor devices. They are mainly used to identify organic specimens—like amino acid chains, genetic material, and even pathogens. In this chapter, we explore the use of biosensor devices, their classifications, and the methods for recognizing the target sample.

There is a wide range of applications for biosensors such as clinical diagnostics, ecological surveillance, pharmaceutical breakthroughs, and bio-manufacturing. Precision is paramount, as is real-time observation and label-independent identification.

The history of optical biosensors is comparatively short when compared with other bio-detection technologies. Initial optical biosensors appeared during the 1960s and 1970s, with the development of optical fiber and laser technology. It was not until the 1980s and 1990s that these biosensors gained increased recognition.

One of the most important developments occurred during the 1980s with the development of surface plasmon resonance (SPR) technology. SPR allows for label-free detection of analytes in real time, making it a valuable tool for protein–protein interactions and ligand–receptor binding studies. This technology led to the creation of commercial optical biosensors, such as Biacore and Reichert SPR instruments [15, 36, 44].

The 1990s saw the emergence of several other optical biosensing technologies, such as evanescent wave spectroscopy, reflectometric interference spectroscopy (RIfS), and fluorescence-based methods. These technologies expanded the range of analytes and applications that could be detected using optical biosensors.

In the early 2000s, researchers began to explore new materials and structures for optical biosensors, such as nanomaterials, plasmonic biosensors, and photonic crystal biosensors. These new developments have allowed for even greater sensitivity and specificity in biosensing applications.

doi:10.1088/978-0-7503-4876-8ch19

19.2 Classification of optical sensors

There are different possibilities for classifying optical biosensors. We can classify the sensors based on the light parameter that we are measuring, like polarization, phase change, intensity, wavelength, and spectral distribution. We could classify them based on their sensing principles like absorption, interference, diffraction, and scattering.

A possible classification for optical biosensors can be based on the optical components where the analyte detection takes place and the detection mechanism. Figure 19.1 shows a bubble map to describe the classifications of biosensors based on parameters mentioned before.

Classifications are useful tools for organizing and comprehending the natural world, but it is essential to recognize that they are inherently limited. Consequently, no classification is perfect, and all such frameworks contain a degree of subjectivity and chance. Consequently, it is difficult to devise a classification scheme capable of precisely encompassing the vast spectrum of optical biosensors and their capabilities.

Figure 19.1. Classification of biosensors based on detection mechanism and structure for analyte detection.

The constant emergence of new biosensor technologies may require the creation of new subcategories or the revision of existing classifications. Consequently, even though optical biosensor classifications are useful for organizing and comprehending the field, they should be viewed as evolving and imperfect representations of a complex and dynamic research domain. In this section, we will analyze optical sensors based on surface plasmon resonance, fluorescence-based sensors, and guided-mode biosensors.

19.2.1 Surface plasmon resonance (SPR) biosensors

SPR sensors are a type of biosensor that use SPR to detect the binding of molecules to a surface [8, 15, 44]. Plasmons are oscillations of free electrons that can be excited at the interface between a metal and a dielectric material. When light is shone on the metal surface at a particular angle, it can couple with the plasmons and create a resonance condition that decreases the reflected light intensity, figure 19.2.

In an SPR sensor, a thin metal film is coated onto a glass or plastic substrate and treated to bind to the target molecule specifically. When a sample containing the target molecule is introduced to the sensor surface, the binding of the target molecule to the metal film leads to a change in the refractive index at the metal–dielectric interface, which results in a shift in the resonance angle and a change in the reflected light intensity [8, 15, 44]. By monitoring these changes, the presence and concentration of the target molecule can be detected.

The thin layer in an SPR sensor is typically a metal film, such as gold or silver, which supports the plasmon oscillations that are used to detect biomolecule binding events. However, the metal film alone may not be suitable for direct binding of biomolecules due to its chemical properties, such as hydrophobicity, and lack of functional groups that can interact with biomolecules [2, 36].

To detect biomolecules, there are several ways to treat the metal film in SPR sensors. Here are a few of them:

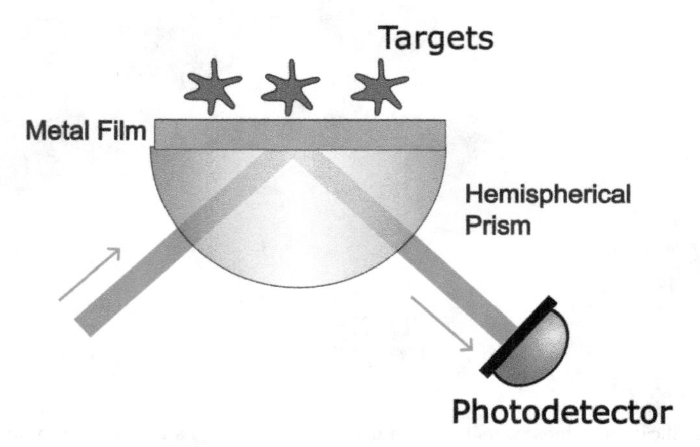

Figure 19.2. Schematic for an SPR sensor.

1. **Self-assembled monolayers (SAMs):** SAMs are a common and versatile surface modification technique involving molecules' adsorption with a thiol group on the metal surface. The thiol group forms a covalent bond with the metal, creating a stable and uniform monolayer of molecules that can be tailored to have specific chemical properties, such as hydrophilicity or charge, to enhance the selectivity and sensitivity of the sensor for certain biomolecules [29, 33, 41].
2. **Covalent immobilization:** covalent immobilization involves attaching a biomolecule, such as an antibody or DNA probe, to the metal surface via a covalent bond. This can be achieved by using a linker molecule, such as a heterobifunctional crosslinker, with a reactive group that can bind to the metal surface and another reactive group that can bind to the biomolecule of interest [27, 41].
3. **Functionalized nanoparticles:** functionalized nanoparticles, such as gold or silver nanoparticles, can be used to modify the metal surface and create a larger surface area for biomolecule binding. The nanoparticles can be functionalized with molecules, such as antibodies or DNA probes, that can specifically bind to the target biomolecule [2, 36].
4. **Polymer brushes:** polymer brushes are a type of surface modification that involve the grafting of polymer chains onto the metal surface. The polymer chains can be designed to have specific chemical properties, such as hydrophilicity or charge, that can enhance the selectivity and sensitivity of the sensor for certain biomolecules [4, 16, 37].

These surface modification techniques can be used to modify the metal film in SPR sensors to improve the sensitivity and selectivity of the sensor for detecting biomolecules. The choice of surface modification technique will depend on factors such as the type of biomolecule being detected, the sensitivity required, and the ease of implementation.

19.2.2 Fluorescence-based biosensors

Fluorescence-based biosensors are a type of biosensors that use fluorescence to detect the presence of a particular molecule or analyte [47]. They use fluorescent proteins, such as green fluorescent protein (GFP), or small organic fluorophores as a detection mechanism.

Fluorescence-based biosensors have many advantages, including high sensitivity and selectivity. This makes them ideal for detecting low concentrations of analytes in complex biological matrices. Fluorescence-based biosensors find applications in diverse fields such as medical diagnosis, drug development, and environmental surveillance [9, 25, 47]. Compared to electrochemical biosensors, they have a higher sensitivity. They can identify different kinds of analytes, including small molecules, proteins, and DNA. Due to their ease of use and miniaturization capability, they are suitable for point-of-care testing.

There are many different types of fluorescence-based biosensors. Some of the most common types include:t

1. **Molecular beacons:** these are small, single-stranded DNA or RNA molecules that are labeled with a fluorophore and a quencher. When the molecular beacon is unbound, the fluorophore and quencher are in close proximity, which quenches the fluorescence. When the molecular beacon binds to its target, the fluorophore and quencher are separated, which allows the fluorescence to be restored. This allows for the sensitive and specific detection of target molecules [25].
2. **FRET-based biosensors:** fluorescence resonance energy transfer (FRET) is a physical phenomenon that occurs over a distance and involves the non-radiative transfer of energy from an excited molecular fluorophore acting as the donor to another fluorophore acting as the acceptor. This transfer is facilitated through intermolecular long-range dipole–dipole coupling. FRET-based biosensors use this phenomenon to detect the binding of a target molecule to a receptor. When the target molecule binds to the receptor, the distance between the donor and acceptor fluorophores is reduced, which allows for FRET to occur. This change in fluorescence can be detected and used to quantify the amount of target molecule present [47].
3. **Fluorescence polarization biosensors:** fluorescence polarization biosensors measure the rotational motion of a fluorophore that is attached to a biomolecule of interest. When the biomolecule is in a certain conformation, the rotational motion of the fluorophore is restricted, resulting in high fluorescence polarization. When the biomolecule changes conformation, the rotational motion of the fluorophore increases, resulting in low fluorescence polarization [5, 9, 14].
4. **Time-resolved fluorescence biosensors:** time-resolved fluorescence biosensors use a fluorophore that emits light after a delay, which allows for the detection of signals with high signal-to-noise ratio. These biosensors can be used to measure binding events and changes in the conformation of the biomolecule of interest [1, 25, 31, 35].

Fluorescence-based biosensors find applications in diverse fields, such as medical diagnosis, drug development, and environmental surveillance. Compared to electrochemical biosensors, they have a higher sensitivity. Different analytes, like small molecules, proteins, and DNA, can be identified through their distinctive characteristics. Due to their ease of use and potential for downsizing, they are highly suitable for the development of portable devices.

Fluorescence-based biosensors have been used to detect a wide range of substances, including glucose, cholesterol, bacteria, viruses, heavy metals, and pesticides [25, 47]. In drug development, they can identify new drugs and monitor drug levels in the body. Furthermore, they are used in environmental surveillance for detecting pollutants in water and air.

19.2.3 Guided-mode biosensors

These biosensors are those in which light propagates inside a structure in defined modes that depend on the dimensions of the structure, the materials on each layer, as well as the light's wavelength. We are considering optical fiber, waveguides, and photonic crystals.

19.2.3.1 Photonic crystals

A photonic crystal is a periodic structure in which the refractive indices vary within the structure. They allow the confinement of light within a limited spatial region and enhance the interaction between light and matter. It is possible to microfabricate semiconductors in order to generate light-manipulating structures using these crystals.

The utilization of this particular structure enables the observation of diverse phenomena that are impossible with conventional materials. These include but are not limited to, robust light confinement, states of decelerated light, and atypical negative refraction phenomena. The utilization of these properties has led to an interesting family of optical sensors.

Various approaches are available for the manufacture of photonic crystals, including electron light beam lithography, and x-ray interference lithography. The process of e-beam lithography involves the utilization of a concentrated electron beam to create perforations in a designated substrate. One benefit of this particular process is that due to the sequential nature of drilling each hole, it is possible to deliberately incorporate flaws into the design by selectively omitting the etching of a specific hole at a given level [40, 46]. Nonetheless, e-beam lithography is associated with two primary drawbacks, namely high cost and slow writing velocities.

X-ray interference lithography is an alternative technique that addresses these issues. Interference lithography utilizes the interference pattern generated by multiple high-frequency beams to imprint a desired pattern onto a photosensitive resin. Following the completion of a specific layer, the resin is subjected to UV light exposure, resulting in its solidification [40, 42]. Subsequently, another layer is added and the procedure is repeated.

This method is not only just a fraction of the cost of e-beam lithography, but is significantly faster, because a single layer can be exposed in just a few seconds. Also due to the small wavelength of the x-ray sources it is possible to have great precision on the design pattern.

One of the drawbacks of x-ray lithography is the complexity involved in determining the suitable beam parameters that can produce the desired interference pattern. Over the past few years, significant progress has been made in overcoming this constraint, successfully implementing various prevalent crystalline configurations [40]. In theory, generating any desired phase pattern is possible by combining phases from an adequate number of beams. However, in practice, this approach is not feasible due to its high level of complexity.

Integrated photonic crystal-based sensors are popular for physical and chemical/biochemical sensing due to their ultra-high light confinement in small volumes, high wavelength selectivity, and ultra-high sensitivity and selectivity in the sensing mechanism [12, 13]. These sensors can be interrogated in two distinct modes: wavelength interrogation mode and intensity interrogation mode. In the wavelength interrogation mode, the optical readout consists of monitoring the wavelength of the optical signal through an optical spectrum analyzer, while in the intensity interrogation mode, the intensity changes of the output signal are monitored using a photodetector.

Refractive index (RI)-based photonic crystal sensors are widely used for detecting chemical and biological species concentrations in gaseous and aqueous environments. These sensors measure RI changes of a bulk solution due to the presence of chemical analytes or gases, which are generally characterized by higher RIs [30, 39]. RI-based photonic crystal sensors offer numerous advantages, such as minimal sample preparation without fluorescence labeling, real-time detection, high sensitivity, and selectivity.

19.2.3.2 Interferometric biosensors

These biosensors are based on the interference of light waves, which can be modulated by the presence of analytes on the sensor surface. Examples include optical waveguide interferometric biosensors and ellipsometric biosensors, which measure changes in light's phase or polarization state upon interaction with the analyte.

In an interferometric biosensor, a beam of light is split into two beams. One beam is directed to a reference surface, and the other beam is directed to a usually, metallic sensing surface. When the target molecules bind to the sensing surface, the dielectric–metal interface changes, and an evanescent field traveling through this interface would be able to interact with the target molecule because the target molecule is placed in the path of the evanescent field. The target molecule absorbs the light in the evanescent field, which changes the RI of the medium. This change in RI can be measured in the interference pattern. Interferometers are very sensitive devices that can detect very small changes in the RI of a medium ($\approx 10^{-7}$). This means that they can be used to detect very low concentrations of target molecules, such toxins, proteins, and viruses [6, 38].

Interferometric biosensors are usually implemented either in optical fibers or waveguides. Some of the possible configurations are shown in figure 19.3. Waveguides and optical fibers are the preferred implementations because they allow bringing the light beam close to the target molecule and concentrating it in order to maximize the interaction between the light and the target to be sensed. The most common configurations used are the Young's and the Mach–Zehnder configurations. These configurations are preferred because of their ability to use a monochromatic light source, they don't have any moving mirrors, and measure only the real part of the index of refraction [6, 28].

Even though optical fibers benefit from the electronics and optical components developed for the telecommunication industry, waveguide optical sensors have the advantage that can be easier fabricated and offer a large option of materials for the substrate and guiding materials (e.g., PMMA, PDMS, silica, glass) [18, 50]. We can

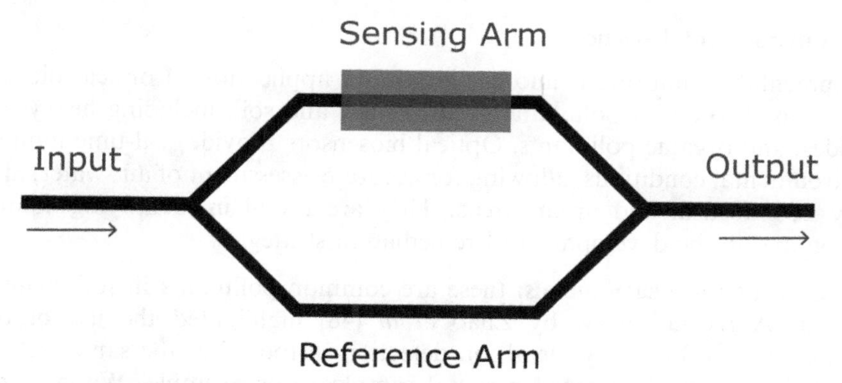

Figure 19.3. Implementation of a Mach–Zehnder interferometer in an optical waveguide.

also customize the geometry of the sensor depending on the application. One of the parameters that we can control when designing a waveguide is the light confinement in the core and the depth-of-field of the evanescent wave.

The evanescent field plays a critical role in determining the sensitivity of an interferometric biosensor. The evanescent field is an electromagnetic field that extends beyond the boundary of the waveguide or optical fiber and decays exponentially with distance from the waveguide or fiber surface. The evanescent field interacts with the surrounding environment and can be used to detect changes in the RI of the environment caused by the presence of target molecules [6, 21, 32]. The binding of a target molecule to the sensing region of a waveguide results in a change in the effective RI of the metal–dielectric structure. As the evanescent wave travels along this interface, it will create a phase difference between the interferometer's two arms. This variation can be perceived as a shift in the output intensity.

The sensitivity of an interferometric biosensor is directly related to the extent of penetration of the evanescent field into the surrounding environment. The penetration depth of the evanescent field depends on several factors, such as the wavelength of the light source, the thickness and RI of the sensing layer, and the angle of incidence of the light, as mentioned in chapter 7. Optimizing these parameters makes it possible to achieve high sensitivity and selectivity for specific target molecules.

19.3 Applications of optical biosensors

Optical biosensors are utilized in a variety of disciplines. Their high sensitivity, real-time monitoring, and label-free detection make them useful instruments for measuring small concentrations of target molecules. In this section, we will explore some of the possible applications of optical biosensors, mainly in the fields of environmental science, defense and homeland security, the food industry, and the medical and health industry. Although this is not an exhaustive list, these are very important fields of application.

19.3.1 Environmental science

Environmental monitoring is another important application of optical biosensors. They are used to detect pollutants in air, water, and soil, including heavy metals, pesticides, and organic pollutants. Optical biosensors provide real-time monitoring of environmental conditions, allowing for accurate assessment of air, water, and soil quality in industrial and urban areas. They are useful in identifying sources of pollution and in the development of remediation strategies.

1. **Detection of heavy metals:** these are common pollutants in soil, water, and air. A recent review by Zhao *et al* [48] highlighted the use of optical biosensors for heavy metal ion detection in food, but the same technology can be applied to environmental samples. For example, Wang *et al* [43] developed an SPR biosensor for the detection of mercury ions in water samples with a limit of detection (LOD) of 0.07 ppb. The biosensor was based on a thymine-rich oligonucleotide sequence that specifically binds to mercury ions, and the SPR signal changes upon binding, allowing for real-time detection.
2. **Detection of pathogens in water:** waterborne pathogens can cause serious health issues, and the traditional methods for detection are time-consuming and labor-intensive. Optical biosensors offer a rapid and sensitive alternative for detecting waterborne pathogens. Yi-Xian *et al* [45] developed a fluorescence-based biosensor for the detection of *Escherichia coli* O157:H7 in water samples. The biosensor was based on the specific binding between the target pathogen and an aptamer probe, which quenches the fluorescence signal. The real-time detection capability of the biosensor allows for early warning of pathogen contamination in water systems.
3. **Detection of pollutants in air:** one example of how optical sensors are used in pollutant detection in the air is the detection of volatile organic compounds (VOCs) using surface-enhanced Raman spectroscopy (SERS)-based optical sensors [11, 23]. VOCs are common air pollutants that can cause respiratory and other health problems. SERS-based optical sensors offer high sensitivity and selectivity for the detection of VOCs in the air.

These examples demonstrate the potential of optical biosensors for environmental monitoring and analysis. Further developments in the field are expected to lead to the commercialization of more sensitive, selective, and cost-effective biosensors for a range of environmental applications.

19.3.2 Food industry

The food industry is another important area where optical biosensors are used extensively. Specifically to evaluate the safety of food sources for human and animal consumption. The main objective of these sensors is to identify pathogens, bacteria, viruses, and other microorganisms that may be harmful if consumed. The high sensitivity and specificity of these detectors makes them valuable in identifying a diverse array of impurities in food. Here are some examples of the use of biosensors in the food industry:

1. **Detection of pathogens:** some of the pathogens that are particularly harmful if consumed are Salmonella, Listeria, and *E. coli*, so particular interest is placed in their detection. An optical biosensor utilizing SPR technology has been devised by researchers, which exhibits remarkable sensitivity and specificity in detecting Salmonella in food samples. The biosensor uses antibodies that specifically bind to Salmonella, causing a change in the SPR signal that can be detected and measured [49]. Optical biosensors have also been used to detect other pathogens, such as viruses and fungi, in food samples [7].
2. **Detection of contaminants:** optical biosensors have also been used to detect contaminants in food, such as pesticides and heavy metals. For example, researchers have developed an optical biosensor based on FRET technology that can detect pesticide residues in food samples with high sensitivity and specificity. The biosensor uses aptamers that specifically bind to the target pesticide, causing a change in the FRET signal that can be detected and measured [17]. Optical biosensors have also been used to detect heavy metals, such as lead and mercury, in food samples [48].
3. **Quality control:** an example of biosensors used in quality control is found in monitoring how fresh is a food product. For example, Endo *et al* [10] developed an optical biosensor that can monitor the freshness of fish in real time. This biosensor uses a fluorophore that emits a signal when exposed to light, but the signal disappears when the fluorophore comes in contact with bacteria that cause spoilage. This allows for real-time monitoring of the freshness of the fish.

19.3.3 Defense and homeland security

Biosensors are essential in safeguarding national security and defense by facilitating the early detection of biological and chemical hazards, as well as monitoring potential infectious disease outbreaks [3, 34]. These devices are designed to rapidly and accurately pinpoint pathogens, toxins, and other agents that could jeopardize public health and safety.

Systems for detecting pathogens are among the most critical applications of biosensors in homeland security [24]. These advanced devices perform a variety of tasks, such as collecting samples, detecting biohazards, preparing samples, and analyzing data. By identifying hazardous agents in the environment, these systems offer early warnings of possible biological or chemical attacks.

The need for quick biosensors that can recognize infectious agents has grown in recent years due to threats from biological terrorism and outbreaks of microbial pathogens [3, 8]. Numerous types of biosensors, including enzyme-based, tissue-based, immunosensors, DNA biosensors, thermal, and piezoelectric biosensors, are being developed and employed.

Besides tracking diseases, discovering drugs, and detecting pollutants, disease-causing microorganisms, and disease indicators in bodily fluids like blood, urine, saliva, and sweat, biosensors are also used for disease monitoring [6, 7].

Biosensors are employed in hazard detection and identification and integrated into planning and response initiatives for bioterrorism incidents. This encompasses creating biological identification systems, improving first responders' technological capabilities, and distributing medical countermeasures in case of an attack [34].

Moreover, surveillance systems have been established to gather and examine data from various sources, such as hospital computer networks, clinical laboratories, electronic health records, and veterinary medical records, to identify potential bioterrorism events as rapidly as possible.

Optical biosensors have several uses in defense, including the detection of biological threats, rapid identification of pathogens, real-time monitoring of biothreats, and point-of-care diagnosis. They are highly sensitive and specific tools that are crucial in ensuring public health and safety.

19.3.4 Health industry

Optical biosensors have various applications in the health industry, including medical diagnostics and drug discovery, due to their benefits, such as low detection limits, high sensitivity, and rapid analysis completion. Optical biosensors significantly impact the health industry as they provide more sensitive, accurate, and rapid diagnostic tools for various diseases.

1. **Medical diagnosis:** optical biosensors are used to detect a wide range of medical conditions such as cancer, infectious diseases, and metabolic disorders [6]. These biosensors detect light properties such as absorbance and fluorescence to convert optical signals into electrical signals. The detection process is based on the interaction of the optical field with a bio-recognition element, which can be categorized as 'label-free' and 'label-based' [7]. The advantages of optical biosensors include ease of use, portability, direct detection, real-time analysis, and inexpensive diagnosis with high sensitivity and specificity [19]. Some of the optical biosensors commonly used in medical diagnosis include colorimetric, fluorometric, luminometric, fiber optic, and SPR-based biosensors.
2. **Drug discoveries:** the sensors allow for the determination of the affinity and kinetics of various molecular interactions in real time, without the need for a molecular tag or label [3]. Optical biosensors such as SPR have been widely used over the past decade to analyze biomolecular interactions and to construct SPR-based sensors that detect biomolecular interactions. The first commercial SPR-based biosensor instrument was developed in the 1990s and is now commonly used in drug discovery processes [2, 15].

   Optical biosensors, particularly label-free technologies, provide high-information content that can enable researchers in the pharmaceutical industry to make better decisions during lead optimization. Optical biosensors such as Biacore T100 and Biacore A100 are commonly used in drug discovery and development to obtain comprehensive information on drug interaction analysis, enabling researchers to make more confident decisions [8]. In addition, the combination of biosensors and nanotechnology has shown intriguing improvements that can be widely used in drug discovery [24].

19.4 Challenges and future perspectives

In recent years, progress in optical biosensors has emphasized the incorporation of new materials and technologies to improve their sensitivity, selectivity, and portability [8]. One such development is the inclusion of nanomaterials in optical biosensors. Gold nanoparticles, for instance, can be utilized as labels to enhance biomolecule detection, while graphene oxide can increase the sensitivity of fluorescence-based biosensors. These nanomaterials can boost the precision and dependability of optical biosensors, expanding their applicability in various fields.

Another significant development is the incorporation of microfluidic systems into optical biosensors [20, 26]. Microfluidic channels can control the flow of samples and reagents, promoting interaction between the sensing element and analyte. This allows real-time monitoring of analytes in complex matrices, such as blood or wastewater. Integrating microfluidics enables optical biosensors to deliver more accurate and precise results more quickly, making them suitable for applications like medical diagnostics and environmental monitoring.

Recent advancements have also been observed in plasmonic biosensors, a type of optical biosensor [2, 4]. Plasmonic biosensors use the interaction between light and metal nanostructures to detect analytes. Current advancements have focused on developing new sensing materials like metal-organic frameworks. By enhancing sensing materials, plasmonic biosensors can provide greater sensitivity and selectivity, making them more effective for detecting low concentrations of analytes.

Smartphone-based biosensors represent another recent development in optical biosensors [22]. These affordable, portable biosensors use a smartphone's camera and processing power to detect and analyze analytes, making them ideal for point-of-care testing. Smartphone-based biosensors have the potential to transform medical diagnostics in resource-limited settings, as they can provide accurate and reliable results without the need for costly laboratory equipment.

Recently, optical biosensor advancements have concentrated on multiplexing, which refers to the simultaneous detection of multiple analytes within a single sample. These multiplexed biosensors can identify and measure numerous analytes in real-time, proving beneficial for medical diagnostics, environmental monitoring, and food safety applications. By increasing the variety of analytes detected in a single sample, multiplexed biosensors offer improved accuracy and efficiency, conserving time and resources across various applications [3].

Optical biosensors hold the potential to bring about significant change in many areas, including medical diagnostics, environmental monitoring, and food safety. One possible future direction involves creating more sensitive and selective sensing materials, such as innovative nanomaterials or advanced plasmonic structures. Such advancements might enable the detection of lower analyte concentrations and the identification of more complex molecules, like proteins or viruses [9, 24].

Another possible direction involves incorporating artificial intelligence (AI) and machine learning algorithms into optical biosensors. AI could analyze vast amounts of biosensor-generated data, yielding more accurate and precise results. Machine learning algorithms might also optimize biosensor design and operation, resulting in enhanced efficiency and decreased costs [18].

Optical biosensors have the potential to become increasingly valuable in detecting and monitoring diseases within medical diagnostics. Future biosensor technology developments could lead to wearable biosensors that monitor patient health in real time. These biosensors could offer continuous vital sign monitoring, such as heart rate and blood glucose levels, resulting in better disease management and improved patient outcomes [6].

In environmental monitoring, optical biosensors could play a crucial role in detecting and monitoring water and air pollutants. Future developments might lead to biosensors capable of simultaneously detecting multiple pollutants, providing a comprehensive view of environmental quality. Optical biosensors could also be integrated into autonomous vehicles or drones for real-time environmental quality monitoring in remote or difficult-to-access locations [18, 47].

In food safety, optical biosensors could become increasingly valuable in detecting pathogens and other contaminants. Future biosensor technology developments could lead to the creation of biosensors capable of detecting a broader range of contaminants and providing faster results. This could help decrease foodborne illness incidence and improve food safety standards [18, 38].

In conclusion, optical biosensors' potential future directions are promising, offering even greater accuracy, reliability, and efficiency. Integrating new materials, AI, and machine learning algorithms could result in more sensitive and selective biosensors, while developing wearable and autonomous biosensors could significantly impact patient health and environmental monitoring. Optical biosensors' role in various fields is likely to grow in the coming years, providing new opportunities to improve human health, environmental quality, and food safety [3, 6, 24].

19.4.1 Limitations

Optical biosensors have some limitations, including:

1. Sensitivity limitations: although optical biosensors are highly sensitive, they may not be able to detect low concentrations of target analytes. This can limit their application in some fields, such as environmental monitoring.
2. Limited selectivity: optical biosensors can sometimes give false positive results due to cross-reactivity with other compounds that are structurally similar to the target analyte. This can reduce their specificity and limit their use in certain applications.

3. Complexity: optical biosensors can be complex and require specialized equipment and skilled operators. This can make them expensive and difficult to use outside of specialized laboratories.
4. Environmental interference: optical biosensors can be susceptible to environmental interference such as temperature changes, pH variations, and electromagnetic interference. This can affect the accuracy of the results and require careful calibration.
5. Limited lifespan: the sensing components in optical biosensors can degrade over time, which can affect their accuracy and sensitivity. This limits the lifespan of the sensors and requires regular maintenance and replacement.

References

[1] Amaro M *et al* 2014 Time-resolved fluorescence in lipid bilayers: selected applications and advantages over steady state *Biophys. J.* **107** 2751–60

[2] Amendola V *et al* 2017 Surface plasmon resonance in gold nanoparticles: a review *J. Phys.: Condens. Matter* **29** 203002

[3] Burcu Bahadr E and Kemal Sezgintürk M 2015 Applications of commercial biosensors in clinical, food, environmental, and biothreat/biowarfare analyses *Anal. Biochem.* **478** 107–20

[4] Balamurugan S *et al* 2003 Thermal response of poly (*N*-isopropylacrylamide) brushes probed by surface plasmon resonance *Langmuir* **19** 2545–9

[5] Benito-Peña E *et al* 2016 Fluorescence based fiber optic and planar waveguide biosensors: a review *Anal. Chim. Acta.* **943** 17–40

[6] Campbell D P 2008 Interferometric biosensors *Principles of Bacterial Detection: Biosensors, Recognition Receptors and Microsystems* ed Zourob M, Elwary S and Turner A (New York: Springer) pp 169–211

[7] Chandra P 2016 *Nanobiosensors for Personalized and Onsite Biomedical Diagnosis (Healthcare Technologies Series)* (London: Institution of Engineering and Technology)

[8] Damborsky P, Švitel J and Katrlík J 2016 Optical biosensors *Essays Biochem.* **60** 91–100

[9] Demchenko A P 2015 *Introduction to Fluorescence Sensing* (Berlin: Springer International Publishing)

[10] Endo H and Wu H 2019 Biosensors for the assessment of fish health: a review *Fish Sci.* **85** 641–54

[11] Fan M, Andrade G F S and Brolo A G 2020 A review on recent advances in the applications of surface-enhanced Raman scattering in analytical chemistry *Anal. Chim. Acta.* **1097** 1–29

[12] Frazão O *et al* 2008 Optical sensing with photonic crystal fibers *Laser Photonics Rev.* **2** 449–59

[13] Hemalatha R and Revathi S 2020 Photonic crystal fiber for sensing application *Int. J. Eng. Adv. Technol.* **9** 481–94

[14] Hendrickson O D *et al* 2020 Fluorescence polarization-based bioassays: new horizons *Sensors* **20** 7132

[15] Homola J 2003 Present and future of surface plasmon resonance biosensors *Anal. Bioanal. Chem.* **377** 528–39

[16] Hu W *et al* 2010 Poly[oligo(ethylene glycol) methacrylate-co-glycidyl methacrylate] brush substrate for sensitive surface plasmon resonance imaging protein arrays *Adv. Funct. Mater.* **20** 3497–503

[17] Kaur J and Singh P K 2020 Enzyme-based optical biosensors for organophosphate class of pesticide detection *Phys. Chem. Chem. Phys.* **22** 15105–19

[18] Li X *et al* 2021 A review of specialty fiber biosensors based on interferometer configuration *J. Biophoton.* **14** e202100068

[19] Liao Y-J and Chen C-C 2015 High-speed CMOS image sensor design for image acquisition systems *J. Electron. Sci. Technol.* **13** 241–6

[20] Liao Z *et al* 2019 Microfluidic chip coupled with optical biosensors for simultaneous detection of multiple analytes: a review *Biosens. Bioelectron.* **126** 697–706

[21] Lin V S-Y *et al* 1997 A porous silicon-based optical interferometric biosensor *Science* **278** 840–3

[22] Lu Y, Shi Z and Liu Q 2019 Smartphone-based biosensors for portable food evaluation *Curr. Opin. Food Sci.* **28** 74–81

[23] Mandal P and Tewari B S 2022 Progress in surface enhanced Raman scattering molecular sensing: a review *Surf. Interfaces* **28** 101655

[24] Mehrotra P 2016 Biosensors and their applications–a review *J. Oral Biol. Craniofac Res.* **6** 153–9

[25] Morris M C 2012 *Fluorescence-Based Biosensors: From Concepts to Applications* (Amsterdam: Elsevier)

[26] Mousavi Shaegh S A *et al* 2016 A microfluidic optical platform for real-time monitoring of pH and oxygen in microfluidic bioreactors and organ-on-chip devices *Biomicrofluidics* **10** 044111

[27] Hiep Nguyen H *et al* 2015 Surface plasmon resonance: a versatile technique for biosensor applications *Sensors* **15** 10481–510

[28] Viet Nguyen L *et al* 2015 Interferometric-type optical biosensor based on exposed core microstructured optical fiber *Sens. Actuators* B **221** 320–7

[29] Okamoto T and Yamaguchi I 2003 Optical absorption study of the surface plasmon resonance in gold nanoparticles immobilized onto a gold substrate by self-assembly technique *J. Phys. Chem.* B **107** 10321–4

[30] Pacholski C 2013 Photonic crystal sensors based on porous silicon *Sensors* **13** 4694–713

[31] Ross J A and Jameson D M 2008 Frequency domain fluorometry: applications to intrinsic protein fluorescence *Photochem. Photobiol. Sci.* **7** 1301–12

[32] Schmitt K *et al* 2007 Interferometric biosensor based on planar optical waveguide sensor chips for label-free detection of surface bound bioreactions *Biosens. Bioelectron.* **22** 2591–7

[33] Siegman A E 1986 *Lasers* (Melville, NY: University Science Books)

[34] Smith R G, D'Souza N and Nicklin S 2008 A review of biosensors and biologically-inspired systems for explosives detection *Analyst* **133** 571

[35] Suhling K *et al* 2007 Time-resolved fluorescence microscopy *SPIE Proc.* **6771** 677106

[36] Tabasi O and Falamaki C 2018 Recent advancements in the methodologies applied for the sensitivity enhancement of surface plasmon resonance sensors *Anal. Methods* **10** 3906–25

[37] Tokareva I *et al* 2004 Nanosensors based on responsive polymer brushes and gold nanoparticle enhanced transmission surface plasmon resonance spectroscopy *J. Am. Chem. Soc.* **126** 15950–1

[38] Tosi D *et al* 2021 *Optical Fiber Biosensors: Device Platforms, Biorecognition, Applications* (Amsterdam: Elsevier)

[39] Troia B *et al* 2013 Photonic crystals for optical sensing: a review *Advances in Photonic Crystals* (InTech Open)

[40] Ullal C K *et al* 2004 Photonic crystals through holographic lithography: simple cubic, diamond-like, and gyroid-like structures *Appl. Phys. Lett.* **84** 5434–6

[41] Unser S *et al* 2015 Localized surface plasmon resonance biosensing: current challenges and approaches *Sensors* **15** 15684–716

[42] Vogelaar L *et al* 2001 Large area photonic crystal slabs for visible light with waveguiding defect structures: fabrication with focused ion beam assisted laser interference lithography *Adv. Mater.* **13** 1551–4

[43] Wang L *et al* 2010 Au NPs-enhanced surface plasmon resonance for sensitive detection of mercury(II) ions *Biosens. Bioelectron.* **25** 2622–6

[44] Wijaya E *et al* 2011 Surface plasmon resonance-based biosensors: from the development of different SPR structures to novel surface functionalization strategies *Curr. Opin. Solid State Mater. Sci.* **15** 208–24

[45] Yi-Xian W *et al* 2012 Application of aptamer based biosensors for detection of pathogenic microorganisms *Chinese J. Anal. Chem.* **40** 634–42

[46] Yablonovitch E 1987 Inhibited spontaneous emission in solid-state physics and electronics *Phys. Rev. Lett.* **58** 2059

[47] Zhang J, Ni Q and Newman R H 2016 *Fluorescent Protein-Based Biosensors: Methods and Protocols* (Totowa, NJ: Humana Press)

[48] Zhao Y *et al* 2020 Optical biosensors for heavy metal ions detection in food: a review *Anal. Chem.* **132** 116040

[49] Zhou J *et al* 2019 Surface plasmon resonance (SPR) biosensors for food allergen detection in food matrices *Biosens. Bioelectron.* **142** 111449

[50] Zinoviev K E *et al* 2011 Integrated bimodal waveguide interferometric biosensor for label-free analysis *J. Light. Technol.* **29** 1926–30

Part IV

Appendices

IOP Publishing

Optical Sensors
An introduction with lab demonstrations
Victor Argueta-Diaz

Appendix A

Vector calculus

Vector calculus is a branch of mathematics that deals with the calculus of vector fields, which are functions that associate a vector to every point in space. Vector calculus includes the study of vector operations such as differentiation, integration, and gradient, as well as the concepts of divergence and curl.

In vector calculus, the gradient is a vector operator that represents the rate of change of a scalar function with respect to its variables. The divergence is a scalar operator that represents the extent to which a vector field originates or terminates at a point, while the curl is a vector operator that represents the rotation of a vector field.

In this appendix, we will present some vector identities that may be helpful when working with electromagnetic waves.

A.1 Vector identities

Let **A** and **B** be vector fields, while U and V are scalar fields.

1. $\nabla \cdot (\mathbf{A} \times \mathbf{B}) = \mathbf{B} \cdot (\nabla \times \mathbf{A}) - \mathbf{A} \cdot (\nabla \times \mathbf{B})$
2. $\nabla \cdot (V\mathbf{A}) = V\nabla \cdot \mathbf{A} + \mathbf{A} \cdot \nabla V$
3. $\nabla \cdot (\nabla \times \mathbf{A}) = 0$
4. $\nabla \times (\nabla V) = 0$
5. $\nabla \times (\mathbf{A} + \mathbf{B}) = \nabla \times \mathbf{A} + \nabla \times \mathbf{B}$
6. $\nabla \times (\mathbf{A} \times \mathbf{B}) = \mathbf{A}(\nabla \cdot \mathbf{B}) - \mathbf{B}(\nabla \cdot \mathbf{A}) + (\mathbf{B} \cdot \nabla)\mathbf{A} - (\mathbf{A} \cdot \nabla)\mathbf{B}$
7. $\nabla \times (\nabla \times \mathbf{A}) = \nabla(\nabla \cdot \mathbf{A}) - \nabla^2\mathbf{A}$
8. $\oint_l \mathbf{A} \cdot d\mathbf{l} = \int_S \nabla \times \mathbf{A} \cdot d\mathbf{S}$
9. $\oint_S \mathbf{A} \cdot d\mathbf{S} = \int_V \nabla \cdot \mathbf{A} d\mathbf{v}$
10. $\oint_S \mathbf{A} \times d\mathbf{S} = -\int_V \nabla \times \mathbf{A} d\mathbf{v}$

doi:10.1088/978-0-7503-4876-8ch20

A.2 Divergence, gradient, curl, and Laplacian

A.2.1 Cartesian coordinates

$$\mathbf{A} = A_x\hat{x} + A_y\hat{y} + A_z\hat{z}$$

$$\nabla V = \frac{\partial V}{\partial x}\hat{x} + \frac{\partial V}{\partial y}\hat{y} + \frac{\partial V}{\partial z}\hat{z}$$

$$\nabla \cdot \mathbf{A} = \frac{\partial A_x}{\partial x}\hat{x} + \frac{\partial A_y}{\partial y}\hat{y} + \frac{\partial A_z}{\partial z}\hat{z}$$

$$\nabla^2 V = \frac{\partial^2 V}{\partial x^2} + \frac{\partial^2 V}{\partial y^2} + \frac{\partial^2 V}{\partial z^2}$$

$$\nabla \times \mathbf{A} = \begin{vmatrix} \hat{x} & \hat{y} & \hat{z} \\ \dfrac{\partial}{\partial x} & \dfrac{\partial}{\partial y} & \dfrac{\partial}{\partial z} \\ A_x & A_y & A_z \end{vmatrix}$$

$$= \left(\frac{\partial A_z}{\partial y} - \frac{\partial A_y}{\partial z}\right)\hat{x} - \left(\frac{\partial A_z}{\partial x} - \frac{\partial A_x}{\partial z}\right)\hat{y} + \left(\frac{\partial A_y}{\partial x} - \frac{\partial A_x}{\partial y}\right)\hat{z}$$

A.2.2 Cylindrical coordinates

$$\mathbf{A} = A_\rho\hat{\rho} + A_\phi\hat{\phi} + A_z\hat{z}$$

$$\nabla V = \frac{\partial V}{\partial \rho}\hat{\rho} + \frac{1}{\rho}\frac{\partial V}{\partial \phi}\hat{\phi} + \frac{\partial V}{\partial z}\hat{z}$$

$$\nabla \cdot \mathbf{A} = \frac{1}{\rho}\frac{\partial}{\partial \rho}\rho A_\rho + \frac{1}{\rho}\frac{\partial A_\phi}{\partial \phi} + \frac{\partial A_z}{\partial z}$$

$$\nabla^2 V = \frac{1}{\rho}\frac{\partial}{\partial \rho}\left(\rho\frac{\partial V}{\partial \rho}\right) + \frac{1}{\rho^2}\frac{\partial^2 V}{\partial \phi^2} + \frac{\partial^2 V}{\partial z^2}$$

$$\nabla \times \mathbf{A} = \frac{1}{\rho}\begin{vmatrix} \hat{\rho} & \rho\hat{\phi} & \hat{z} \\ \dfrac{\partial}{\partial\rho} & \dfrac{\partial}{\partial\phi} & \dfrac{\partial}{\partial z} \\ A_\rho & \rho A_\phi & A_z \end{vmatrix}$$

$$= \left(\frac{1}{\rho}\frac{\partial A_z}{\partial\phi} - \frac{\partial A_\phi}{\partial z}\right)\hat{\rho} - \left(\frac{\partial A_z}{\partial\rho} - \frac{\partial A_\rho}{\partial z}\right)\hat{\phi} + \frac{1}{\rho}\left(\frac{\rho\partial A_\phi}{\partial\rho} - \frac{\partial A_\rho}{\partial\phi}\right)\hat{z}$$

A.2.3 Spherical coordinates

$$\mathbf{A} = A_r\hat{r} + A_\theta\hat{\theta} + A_\phi\hat{\phi}$$

$$\nabla V = \frac{\partial V}{\partial r}\hat{r} + \frac{1}{r}\frac{\partial V}{\partial\theta}\hat{\theta} + \frac{1}{r\sin\theta}\frac{\partial V}{\partial\phi}\hat{\phi}$$

$$\nabla \cdot \mathbf{A} = \frac{1}{r^2}\frac{\partial}{\partial r}r^2 A_r + \frac{1}{\rho}\frac{\partial A_\phi}{\partial\phi} + \frac{\partial A_z}{\partial z}$$

$$\nabla^2 V = \frac{1}{r^2}\frac{\partial}{\partial r}\left(r^2\frac{\partial V}{\partial r}\right) + \frac{1}{r^2\sin\theta}\frac{\partial}{\partial\theta}\left(\sin\theta\frac{\partial V}{\partial\theta}\right) + \frac{1}{r^2\sin^2\theta}\frac{\partial^2 V}{\partial\phi^2}$$

$$\nabla \times \mathbf{A} = \frac{1}{r^2\sin\theta}\begin{vmatrix} \hat{r} & r\hat{\theta} & (r\sin\theta)\hat{\phi} \\ \dfrac{\partial}{\partial r} & \dfrac{\partial}{\partial\theta} & \dfrac{\partial}{\partial\phi} \\ A_r & rA_\theta & (r\sin\theta)A_\phi \end{vmatrix}$$

$$= \frac{1}{r\sin\theta}\left(\frac{\partial}{\partial\theta}(A_\phi\sin\theta) - \frac{\partial A_\theta}{\partial\phi}\right)\hat{r} - \frac{1}{r}\left(\frac{\partial}{\partial r}rA_\phi - \frac{1}{\sin\theta}\frac{\partial A_r}{\partial\phi}\right)\hat{\theta} + \frac{1}{r}\left(\frac{\partial rA_\theta}{\partial r} - \frac{\partial A_r}{\partial\theta}\right)\hat{\phi}$$

Appendix B

Fields in waveguides and optical fibers

B.1 Electric and magnetic fields in a dielectric rectangular waveguide

Assuming a general rectangular waveguide structure such as the one shown in figure B.1. We can find the electric and magnetic fields for an E^x-mode in the non-shadow regions from their longitudinal components E_z, and H_z. For all the following fields, we are assuming a z-propagation factor $\exp(-j\beta z)$

Region 1:

$$E_z = A_0 \cos(\kappa_x(x + \phi_x))\cos(\kappa_y(y + \phi_y))$$

$$E_x = \frac{jA_0}{\kappa_x\beta}(k_0^2 n_1^2 - \kappa_x^2)\sin(\kappa_x(x + \phi_x))\cos(\kappa_y(y + \phi_y))$$

$$E_y = -\frac{jA_0\,\kappa_1}{\beta}\cos(\kappa_x(x + \phi_x))\sin(\kappa_y(y + \phi_y))$$

$$H_z = -A_0 n_1^2\sqrt{\frac{\epsilon_0}{\mu_0}}\frac{\kappa_y k_0}{\kappa_x\beta}\sin(\kappa_x(x + \phi_x))\sin(\kappa_y(y + \phi_y))$$

$$H_x = 0$$

$$H_y = -A_0 n_1^2\sqrt{\frac{\epsilon_0}{\mu_0}}\frac{\kappa_y k_0}{\kappa_x\beta}\sin(\kappa_x(x + \phi_x))\sin(\kappa_y(y + \phi_y))$$

doi:10.1088/978-0-7503-4876-8ch21

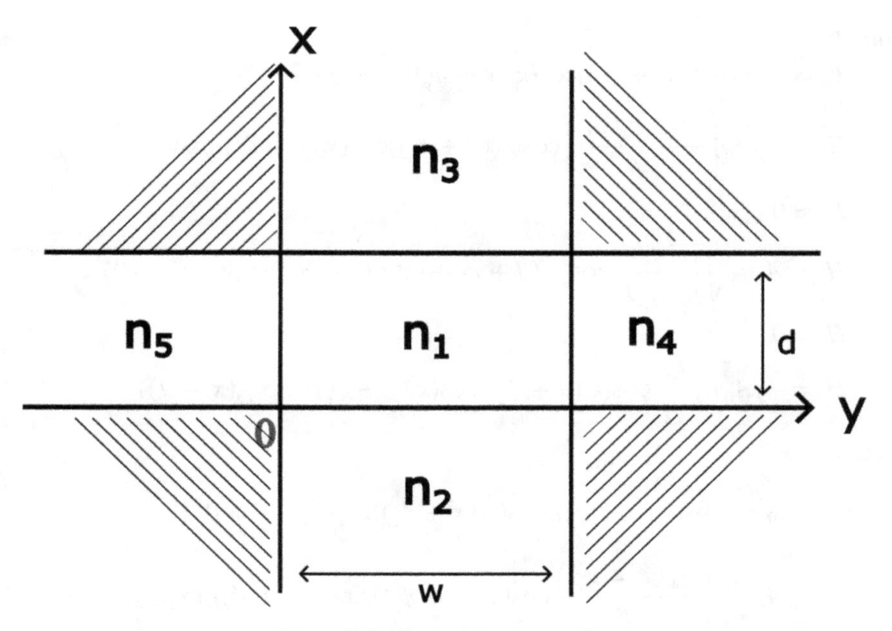

Figure B.1. Rectangular waveguide.

Region 2:

$$E_z = A_0 \cos(\kappa_x \phi_x)\cos(\kappa_y(y + \phi_y))\exp(\gamma_2 x)$$

$$E_x = jA_0\left(\frac{\gamma_2^2 + n_2^2 k_0^2}{\gamma_2 \beta}\right)\cos(\kappa_x \phi_x)\cos(\kappa_y(y + \phi_y))\exp(\gamma_2 x)$$

$$E_y \approx 0$$

$$H_z = -A_0 n_2^2 \sqrt{\frac{\epsilon_0}{\mu_0}}\frac{\kappa_y k_0}{\gamma_2 \beta}\cos(\kappa_x \phi_x)\cos(\kappa_y(y + \phi_y))\exp(\gamma_2 x)$$

$$H_x = 0$$

$$H_y = jA_0 n_2^2 \frac{k_0}{\gamma_2}\cos(\kappa_x \phi_x)\cos(\kappa_y(y + \phi_y))\exp(\gamma_2 x)$$

Region 3:

$$E_z = A_0 \cos(\kappa_x(d + \phi_x))\cos(\kappa_y(y + \phi_y))\exp(\gamma_3(x - d))$$

$$E_x = -jA_0\left(\frac{\gamma_3^2 + n_3^2k_0^2}{\gamma_3\beta}\right)\cos(\kappa_x(d + \phi_x))\cos(\kappa_y(y + \phi_y))\exp(\gamma_3(x - d))$$

$$E_y \approx 0$$

$$H_z = A_0 n_3^2\sqrt{\frac{\epsilon_0}{\mu_0}}\frac{\kappa_y k_0}{\gamma_3\beta}\cos(\kappa_x(d + \phi_x))\sin(\kappa_y(y + \phi_y))\exp(\gamma_3(x - d))$$

$$H_x = 0$$

$$H_y = -jA_0 n_3^2\frac{k_0}{\gamma_3}\cos(\kappa_x(d + \phi_x))\cos(\kappa_y(y + \phi_y))\exp(\gamma_3(x - d))$$

Region 4:

$$E_z = A_0\frac{n_1^2}{n_4^2}\cos(\kappa_y(w + \phi_y))\cos(\kappa_x(x + \phi_x))\exp(\gamma_4(y - w))$$

$$E_x = jA_0\frac{n_1^2}{n_4^2}\left(\frac{n_4^2k_0^2 - \kappa_x^2}{\kappa_x\beta}\right)\cos(\kappa_y(w + \phi_y))\sin(\kappa_x(x + \phi_x))\exp(\gamma_4(y - w))$$

$$E_y \approx 0$$

$$H_z = -A_0 n_1^2\sqrt{\frac{\epsilon_0}{\mu_0}}\frac{\gamma_4 k_0}{\kappa_x\beta}\cos(\kappa_y(w + \phi_x))\sin(\kappa_x(x + \phi_x))\exp(\gamma_4(y - w))$$

$$H_x = 0$$

$$H_y = jA_0 n_1^2\sqrt{\frac{\epsilon_0}{\mu_0}}\frac{k_0}{\kappa_x}\cos(\kappa_y(w + \phi_y))\sin(\kappa_x(x + \phi_x))\exp(\gamma_4(y - w))$$

Region 5:

$$E_z = A_0\frac{n_1^2}{n_5^2}\cos(\kappa_y\phi_y)\cos(\kappa_x(x + \phi_x))\exp(\gamma_5 y)$$

$$E_x = jA_0\frac{n_1^2}{n_5^2}\left(\frac{n_5^2k_0^2 - \kappa_x^2}{\kappa_x\beta}\right)\cos(\kappa_y\phi_y)\sin(\kappa_x(x + \phi_x))\exp(\gamma_5 y)$$

$$E_y \approx 0$$

$$H_z = A_0 n_1^2\sqrt{\frac{\epsilon_0}{\mu_0}}\frac{k_0}{\kappa_x}\cos(\kappa_y\phi_x)\sin(\kappa_x(x + \phi_x))\exp(\gamma_y y)$$

$$H_x = 0$$

$$H_y = jA_0 n_1^2\sqrt{\frac{\epsilon_0}{\mu_0}}\frac{k_0}{\kappa_x}\cos(\kappa_y\phi_y)\sin(\kappa_x(x + \phi_x))\exp(\gamma_5 y)$$

B.2 Electric and magnetic fields in an optical fiber

B.2.1 TE modes

For TE modes we have, $E_r = E_z = H_\theta = 0$

For fields in the core $r \leqslant a$:

$$E_\theta = -j\frac{\omega\mu_0}{\kappa}A_0 J_1(\kappa r)$$

$$H_r = j\frac{\beta}{\kappa}A_0 J_1(\kappa r)$$

$$H_z = A_0 J_0(\kappa r)$$

For fields in the cladding $r \geqslant a$:

$$E_\theta = j\frac{\omega\mu_0}{\gamma}\frac{J_0(\kappa a)}{K_0(\gamma a)}A_0 K_1(\gamma r)$$

$$H_r = -j\frac{\beta}{\kappa}\frac{J_0(\kappa a)}{K_0(\gamma a)}A_0 K_1(\gamma r)$$

$$H_z = A_0\frac{J_0(\kappa a)}{K_0(\gamma a)}K_0(\gamma r)$$

B.2.2 TM modes

For TM modes we have, $H_r = H_z = E_\theta = 0$

For fields in the core $r \leqslant a$:

$$E_r = j\frac{\beta}{\kappa}A_0 J_1(\kappa r)$$

$$E_z = A_0 J_0(\kappa r)$$

$$H_\theta = j\omega\epsilon_0 n_1^2 A_0 J_1(\kappa r)$$

For fields in the cladding $r \geqslant a$:

$$E_r = j\frac{\omega\mu_0}{\gamma}\frac{J_0(\kappa a)}{K_0(\gamma a)}A_0 K_1(\gamma r)$$

$$E_z = A_0\frac{J_0(\kappa a)}{K_0(\gamma a)}K_0(\gamma r)$$

$$H_\theta = -j\frac{n_2^2\omega\epsilon_0}{\gamma}\frac{J_0(\kappa a)}{K_0(\gamma a)}A_0 K_1(\gamma r)$$

B.2.3 Hybrid modes

In hybrid modes we have non-zero E_z and H_z. Solutions are given by the product of Bessel functions and $\cos(n\theta + \phi)$ or $\sin(n\theta + \phi)$

For fields in the core $r \leqslant a$:

$$E_r = -jA_0 \frac{\beta}{\kappa}\left[\frac{1-s}{2}J_{n-1}(\kappa r) - \frac{1+s}{2}J_{n+1}(\kappa r)\right]\cos(n\theta + \phi)$$

$$E_\theta = jA_0 \frac{\beta}{\kappa}\left[\frac{1-s}{2}J_{n-1}(\kappa r) - \frac{1+s}{2}J_{n+1}(\kappa r)\right]\sin(n\theta + \phi)$$

$$E_z = AJ_n(\kappa r)\cos(n\theta + \phi)$$

$$H_r = -jA_0 \frac{\omega n_1^2}{\kappa}\left[\frac{1-s_1}{2}J_{n-1}(\kappa r) + \frac{1+s_1}{2}J_{n+1}(\kappa r)\right]\sin(n\theta + \phi)$$

$$H_\theta = -jA_0 \frac{\omega n_1^2}{\kappa}\left[\frac{1-s_1}{2}J_{n-1}(\kappa r) - \frac{1+s_1}{2}J_{n+1}(\kappa r)\right]\cos(n\theta + \phi)$$

$$H_z = -A_0 \frac{\beta s}{\omega \mu_0} J_n(\kappa r)\sin(n\theta + \phi)$$

For fields in the cladding $r \geqslant a$:

$$E_r = -jA_0 \frac{\beta}{\gamma}\frac{J_n(\kappa a)}{K_n(\gamma a)}\left[\frac{1-s}{2}K_{n-1}(\gamma r) + \frac{1+s}{2}K_{n+1}(\gamma r)\right]\cos(n\theta + \phi)$$

$$E_\theta = jA_0 \frac{\beta}{\gamma}\frac{J_n(\kappa a)}{K_n(\gamma a)}\left[\frac{1-s}{2}K_{n-1}(\gamma r) - \frac{1+s}{2}K_{n+1}(\gamma r)\right]\sin(n\theta + \phi)$$

$$E_z = A_0 \frac{J_n(\kappa a)}{K_n(\gamma a)} K_n(\gamma r)\cos(n\theta + \phi)$$

$$H_r = -jA_0 \frac{\omega \epsilon_0 n_2^2}{\gamma}\frac{J_n(\kappa a)}{K_n(\gamma a)}\left[\frac{1-s_2}{2}K_{n-1}(\gamma r) - \frac{1+s_2}{2}K_{n+1}(\gamma r)\right]\sin(n\theta + \phi)$$

$$H_\theta = -jA_0 \frac{\omega \epsilon_0 n_2^2}{\gamma}\frac{J_n(\kappa a)}{K_n(\gamma a)}\left[\frac{1-s_2}{2}K_{n-1}(\gamma r) + \frac{1+s_2}{2}K_{n+1}(\gamma r)\right]\cos(n\theta + \phi)$$

$$H_z = -A_0 \frac{\beta s}{\omega \mu_0}\frac{J_n(\kappa a)}{K_n(\gamma a)} K_n(\gamma r)\sin(n\theta + \phi)$$

where:

$$s = \frac{n\left(\frac{1}{(\kappa a)^2} + \frac{1}{(\gamma a)^2}\right)}{\left[\frac{J_n'(\kappa a)}{\kappa a J_n(\kappa a)} + \frac{K_n'(\gamma a)}{\gamma a K_n(\gamma a)}\right]}$$

$$s_1 = \frac{\beta^2}{k_0^2 n_1^2} s$$

and,

$$s_2 = \frac{\beta^2}{k_0^2 n_2^2} s$$

Appendix C

Useful constants

Speed of light in a vacuum, $c = 299\ 792\ 458\ \mathrm{m\ s^{-1}}$

Permittivity of free space, $\epsilon_0 = 8.854\ 187\ 82 \times 10^{-12}\ \mathrm{F\ m^{-1}}$

Permeability of free space, $\mu_0 = 4\pi \times 10^{-7}\mathrm{H\ m^{-1}}$

Free-space impedance, $Z_0 = 376.730\ 313\ \Omega$

Planck constant, $h = 6.626\ 070\ 15 \times 10^{-34}\ \mathrm{J\ s}$

Electron charge, $e = 1.602\ 176\ 634 \times 10^{-19}\ \mathrm{C}$

Electron mass, $m_\mathrm{e} = 9.1094 \times 10^{-28}\ \mathrm{g}$

Protonmass, $1.6726 \times 10^{-24}\ \mathrm{g}$

Neutronmass, $1.6749 \times 10^{-24}\ \mathrm{g}$

doi:10.1088/978-0-7503-4876-8ch22

Printed in the USA
CPSIA information can be obtained
at www.ICGtesting.com
JSHW060712031123
51216JS00004B/94